ACKNOWLEDGEMENTS

This report was prepared on the basis of a mission to Nepal in November 1989 consisting of W. James Smith (Mission Leader), John H. Duloy (Consultant - Poverty and Incomes), Christopher Gibbs (Agricultural Economist), Lynn Bennett (Anthropologist), John Elder (Consultant - Agriculture, Food Security), Sanjay Sinha (Consultant - Informal Sector) and Neil Walton (Consultant - NGO and Income-Generating Programs).

The analysis is based on a substantial amount of preparatory work that was done in Nepal; background studies were managed by Meena Acharya, Kul Shekhar Sharma, Madhup Dhungana and Harka Gurung. Thanks are due to the Nepal Rastra Bank for allowing us access to the Multi-Purpose Household Budget Survey Data, and particular recognition is due to Baskar Risal, who managed the reprocessing of the survey data.

The report draws on numerous background papers and analyses, to which the following persons contributed: Messrs./Mss. Ram Nath Acharya, Hamid Ansari, Aswasthama, Kishor Kumar Guru Gharana, Ana Maria Jeria, Chaitanya Mishra, Yogendra Prasai, Khem Raj Sharma, Azealia Ranjitkar, Prayograj Sharma, Ramesh Sharma, Salikram Sharma, Gajendra Man Shrestha, Subrama Laxmi Singh, Ganesh Bahadur Thapa, Indra Jung Thapa, and William Thiesenhusen. Irajen Appasamy provided computational support, and Deborah Ricks processed the report.

Financial support for the study was provided by the United Nations Development Program. The report also benefited from the findings of a parallel study of poverty undertaken by the Canadian International Development Agency.

A WORLD BANK COUNTRY STUDY

Nepal
Poverty and Incomes

A Joint Study

The World Bank

The United Nations
Development Programme

The World Bank
Washington, D.C.

HC
425
Z9
P664
1991

Copyright © 1991
The International Bank for Reconstruction
and Development/THE WORLD BANK
1818 H Street, N.W.
Washington, D.C. 20433, U.S.A.

40702

All rights reserved
Manufactured in the United States of America
First printing April 1991

World Bank Country Studies are among the many reports originally prepared for internal use as part of the continuing analysis by the Bank of the economic and related conditions of its developing member countries and of its dialogues with the governments. Some of the reports are published in this series with the least possible delay for the use of governments and the academic, business and financial, and development communities. The typescript of this paper therefore has not been prepared in accordance with the procedures appropriate to formal printed texts, and the World Bank accepts no responsibility for errors.

The World Bank does not guarantee the accuracy of the data included in this publication and accepts no responsibility whatsoever for any consequence of their use. Any maps that accompany the text have been prepared solely for the convenience of readers; the designations and presentation of material in them do not imply the expression of any opinion whatsoever on the part of the World Bank, its affiliates, or its Board or member countries concerning the legal status of any country, territory, city, or area or of the authorities thereof or concerning the delimitation of its boundaries or its national affiliation.

The material in this publication is copyrighted. Requests for permission to reproduce portions of it should be sent to Director, Publications Department, at the address shown in the copyright notice above. The World Bank encourages dissemination of its work and will normally give permission promptly and, when the reproduction is for noncommercial purposes, without asking a fee. Permission to photocopy portions for classroom use is not required, though notification of such use having been made will be appreciated.

The complete backlist of publications from the World Bank is shown in the annual *Index of Publications*, which contains an alphabetical title list (with full ordering information) and indexes of subjects, authors, and countries and regions. The latest edition is available free of charge from the Publications Sales Unit, Department F, The World Bank, 1818 H Street, N.W., Washington, D.C. 20433, U.S.A., or from Publications, The World Bank, 66, avenue d'Iéna, 75116 Paris, France.

ISSN: 0253-2123

Library of Congress Cataloging-in-Publication Data

```
Nepal : poverty and incomes.
      p.   cm. -- (A World Bank country study)
   "A joint study by the World Bank and the United Nations
Development Programme."
   ISBN 0-8213-1808-X
   1. Poor--Nepal.  2. Income distribution--Nepal.  3. Nepal-
-Economic conditions.  4. Nepal--Social conditions.  5. Nepal-
-Economic policy.  6. Nepal--Social policy.  I. International Bank
for Reconstruction and Development.  II. United Nations Development
Programme.  III. Series.
HC425.Z9P664   1991
362.5'095496--dc20                                         91-12947
                                                               CIP
```

CURRENCY EQUIVALENTS

Year	US$1.00 Equivalent (Average)
1984/85	Rs. 17.8
1988/89	Rs. 25.6

FISCAL YEAR

July 16 - July 15

LIST OF ABBREVIATIONS AND ACRONYMS USED

ADB	Asian Development Bank
ADB/N	Agricultural Development Bank of Nepal
AIC	Agricultural Inputs Corporation
BNP	Basic Needs Program
CSI	Cottage and Small Industries
EGS	Employment Guarantee Scheme
ERL	Environmental Resources Ltd.
ha.	hectare
HMG	His Majesty's Government of Nepal
IBP	Intensive Banking Program
IRDP	Integrated Rural Development Project
kg.	kilogram
LF	Labour Force
MPHBS	Multi Purpose Household Budget Survey
mt.	metric tons
NFC	National Food Corporation
NGO	Non-Governmental Organization
NPC	National Planning Commission
p.a.	per annum
PCRW	Production Credit for Rural Women
Rs.	Rupees
SFDP	Small Farmers' Development Program
SSSR	Social Sector Strategy Review
UNDP	United Nations Development Program
UNICEF	United Nations Children's Fund
WDR	World Development Report
WFP	World Food Program

Table of Contents

		Page No.
	Executive Summary......................................	xi
CHAPTER I:	Introduction...	1
	A. Objectives and Scope............................	1
	B. The Nature of the Problem........................	3
CHAPTER II:	The Structure of Incomes and Poverty.................	5
	A. Levels and Distribution of Income in Nepal........	5
	B. Composition of Income............................	9
	C. Profile of the Poor.............................	13
	D. Conditions of the Poor...........................	16
CHAPTER III:	The Determinants of Poverty..........................	23
	A. The Dynamics of Poverty- Output and Growth Issues.................................	23
	Population and the Resource Base................	23
	Economic Performance............................	26
	Macroeconomic Conditions........................	27
	Structural Adjustment...........................	29
	The Trade and Transit Dispute...................	30
	Emerging Trends.................................	30
	B. The Mechanics of Poverty - Distributional and Equity Issues..............................	33
	The Social Context - Inequality at the Village Level................................	33
	Land Tenure and Tenancy.........................	34
	Labour Force Issues.............................	38
	Migration.......................................	45
	Debt and Indebtedness...........................	46
	Gender and Poverty..............................	48
	Social and Political Constraints................	50
CHAPTER IV:	Poverty and the Productive Sectors....................	53
	A. Agricultural Incomes and Poverty.................	53
	Poverty and the Environment.....................	64
	B. Off-Farm Employment and Incomes..................	65
	The Formal Sector...............................	65
	The Informal Sector.............................	69

CHAPTER V:	The Social Dimensions of Poverty............................	73
	A. Nutrition and Access to Food......................	73
	B. Education and the Poor...........................	78
	C. Poverty, Population and Health...................	85
CHAPTER VI:	Poverty Related Programs and Policies..................	91
	A. HMG's Policies and Programs......................	91
	General Policies................................	91
	The Composition of Public Expenditure...........	91
	The Basic Needs Program and the Eighth Plan......	93
	B. Poverty Alleviation Programs.....................	94
	Subsidies and Transfers.........................	94
	Integrated Rural Development Projects...........	97
	Food and Feeding Programs.......................	99
	Targetted Credit................................	101
	Employment Creation Projects....................	105
	Income Generating Projects......................	106
	C. Institutional Issues.............................	107
	Civil Service and Service Delivery Issues.......	107
	Decentralization and Popular Participation......	108
	Non-Government Organizations....................	109
CHAPTER VII:	Country Strategy Implications...........................	111
CHAPTER VIII:	Policy Conclusions.....................................	119
	A. Overall Strategy.................................	119
	B. Sectoral Policies................................	120
	C. Programming Implications........................	124
ANNEX I:	Summary of Detailed Recommendations.....................	127
ANNEX II:	Statistical Appendices..................................	135
ANNEX III:	Bibliography..	215

Map No. IBRD 22886

List of Tables

Table 1.1:	Indicators of Relative Underdevelopment-Selected Countries..	3
Table 1.2:	Structural Economic Indicators..........................	4
Table 2.1:	Average Per Capita Monthly Income by Income Deciles....	5
Table 2.2:	Distribution of Total Household Income in Nepal and Selected Countries.................................	6
Table 2.3:	Per Capita Household Income Distribution in Nepal......	7
Table 2.4:	Incidence of Poverty Under Different Poverty Lines.....	8
Table 2.5:	Composition of Household Income-1984/85.................	10
Table 2.6:	Some Survey Estimates of Income Composition............	11
Table 2.7:	The Effect of Farm Size on Incomes.....................	12
Table 2.8:	Returns to Family Labor Per Day Worked By Farm Size....	13
Table 2.9:	Nepal - Poverty Profile................................	14
Table 2.10:	Monthly Consumption Pattern of Poor Households.........	17
Table 2.11:	Food Sources and Consumption Among the Poor-Rural Terai..	18
Table 3.1:	GDP Composition and Growth.............................	26
Table 3.2:	Nepal - Structural Economic Indicators.................	28
Table 3.3:	Size Distribution of Operational Holdings..............	34
Table 3.4:	Estimated Current Daily Wage Rates (1989)..............	40
Table 4.1:	Characteristics of Farm Household Groups in the Hills....................................	57
Table 4.2:	Potential Monthly Per Capita Income From Field Crops and Percentage of Poverty Line Income For Poor Households, 1988/89.....................................	59
Table 4.3:	Household Income and Landholdings in the Terai by Level of Poverty.....................................	60
Table 4.4:	Potential Income Per Hectare from Field Crops in the Terai and Area Needed for Poverty Line Income, 1988/89..	61
Table 4.5:	Current and Projected Formal Sector Employment.........	68

Table 4.6:	Estimated Informal Sector Employment - 1990	69
Table 5.1:	Regional Per Capita Production of Grains and Potatoes	75
Table 5.2:	Population Growth Rates (1971-1981) and Growth Rates of Agricultural Production (1967/68-1987/88)	76
Table 5.3:	Projected Daily Per Capita Availability of Calories from Grains and Potatoes	77
Table 6.1:	Composition of Public Expenditure - 1988/89	92
Table 6.2:	Nepal - Poverty Alleviation Programs at a Glance	94
Table 6.3:	Targetted Credit Programs	102
Table 7.1:	Possible Labour Absorption - 2010	113
Table 7.2:	Potential Impact of Various Developments on the Incidence of Poverty	114

List of Figures

Figure 1:	Composition of Income Among the Rural Poor	9
Figure 2:	Effect of Population Growth on Income and Food Availability	24
Figure 3:	Actual and Projected Population Growth 1910-2030	30
Figure 4:	The Changing Spatial Distribution of Population	31
Figure 5:	Annual and Trend Variation in Food Production	76
Figure 6:	Income, Gender, and Education Participation	79

Executive Summary

1. This study seeks to build a better understanding of the nature of incomes and poverty in Nepal, and to propose an affordable set of measures to reduce the incidence of poverty. It consists of three main elements:

- a quantitative analysis of the conditions of the poor and the non-poor, their sources and levels of income;

- an analysis of the micro-economic factors influencing incomes, the constraints facing the poor, and the potential contribution of various sectors to raising personal incomes; and,

- an appraisal of the effectiveness of existing poverty alleviation programs and projects.

It concludes by outlining the elements of a medium- to long-term poverty alleviation strategy for Nepal.

2. By the most conservative definition, between 7 and 8 million of Nepal's population of 19 million live in absolute poverty, defined as having incomes below the level required to support a minimum daily calorie intake (about US$100 p.a. per capita). The poor are overwhelmingly rural subsistence farmers. They earn about half their incomes from their own agricultural production (almost none of which is marketed), they earn less than 30% from employment - mostly on-farm, and the remainder of their incomes is made up from miscellaneous subsistence activities. Although there has been some shift towards off-farm activities, the limited resource base, and lack of alternative opportunities resulting from Nepal's landlocked position beside an economically dominant neighbor, have limited non-agricultural employment growth.

3. Population growth of 2.7% p.a. has eroded the limited gains that have been made in GDP and agricultural output. The population has doubled since 1960, and is projected to double again over the next 25 years. Furthermore, the demographic profile is such that within ten years the labour force will be growing at about 0.4 million persons per year - twice the average rate experienced during the 1980's. The major challenge in avoiding a deterioration in the poverty situation will be managing the absorption of this massive labour force growth. However, in the absence of an effective program to slow population growth, all other poverty-alleviation measures will be meaningless.

4. At the very low level of average GDP, raising personal incomes for most Nepalese depends on overall economic growth. Given the limited cultivable land base, agriculture alone cannot ultimately be counted on to provide the solution to poverty in Nepal. The basis for long-term growth, if it is to come at all, must thus eventually be sought in the expansion of services, energy, and industry. For these it seems inevitable that Nepal

will have to look to a large extent to greater participation in a growing Indian economy. However even with the best policies and most robust external environment, industrial sector growth will be a very long term proposition. Therefore, in the medium-term raising incomes will have to rely in large part on agricultural intensification, and agriculturally-led growth in the informal sector.

5. The agricultural land base is rapidly approaching saturation. There is, however, scope for increases in both labour absorption and agricultural productivity - largely through improved irrigation. About half of the agricultural poor could rise out of poverty on the basis of increased productivity, however to reach them will require a more subtle blend of agricultural interventions than has been tried to date. For the balance, their holdings are too small to ever be viable economic units - they need off-farm income-earning opportunities - either where they currently are, or elsewhere.

6. Even under the most optimistic assumptions, the formal sector will not absorb more than about 15-20% of the labor force by 2010. The informal sector holds more promise, since there is room for a "catch-up" effect to compensate for the low levels of physical access and monetization in the past; although ultimately it can only follow growth led by the other sectors - especially agriculture. However, wages are close to a subsistence minimum, and at existing wage levels the poor will not rise out of poverty on the basis of employment alone - that will require a tightening of the labour market. While a successful population program can contribute, in the long run the scope for effecting labour market factors may be limited by the free flow of labour to and from India, which will tend to lead to wages equalizing at Indian levels. Given the constraints on effective labour market interventions, Government can best contribute to off-farm employment growth through enabling mechanisms: providing transport and communications infrastructure, education, and to a lesser extent skills training and credit.

7. In the absence of any obvious source of rapid economic expansion, curbing population growth is central to relieving poverty. However, at any expected levels of population and GDP growth, average incomes will not rise enough to have a major impact on poverty levels _unless_ such growth is focussed to benefit the poor. Furthermore, forces are in place - through growing landlessness, monetization, and urbanization - for a deterioration in the distribution of income, unless policies are followed which consciously guard against it. The type of growth pursued must be balanced in such a way as to generate incomes for the poor (in the medium term this means largely increased agricultural productivity and labor absorption).

8. Poverty in Nepal is chronic, basically rooted in the insufficiency of the resource base vis. a vis. excessive population. Its solution will lie in productivity growth coupled with population control, but this will take a long time. In the meantime, there will remain a large number of absolute poor. It is therefore legitimate to consider a sustained program of support to the poor - some of it production-oriented and some of it welfare-oriented.

9.   There are too many poor (and too few resources) to realistically consider large-scale transfer or subsidy programs. Therefore it is important to design cost-effective transfers. Both Government and the donors, under a wide-range of projects, are already putting substantial resources into poverty-type programs - to relatively little effect. Better targetting and institutional strengthening are needed to improve the efficiency with which those resources are used. In addition, some significant welfare improvements for the poor can be achieved without major resource transfers or income increases (through, for example, improved hygiene and nutrient retention).

10.   The report concludes that there is no easy poverty alleviation strategy for Nepal, but that significant gains can be made through a combination of measures - mostly involving increased labour absorption in agriculture (which could increase by about 50%), coupled with productivity gains in low-input farming systems; informal sector growth; and some redistributive measures, if tightly focussed.

11.   There are also some areas (eg., tenancy, labor contracting) where policy reforms are required. In addition, limited gains can be made at the micro level, by strengthening the capacity of the poor to undertake self-reliant income generation - although these efforts require labor-intensive inputs and their replicability is thus questionable. Given the weakness of service delivery mechanisms these are probably best delivered through non-governmental organizations. While the report concludes that credit and income-generating projects are unlikely to have a large-scale impact on incomes of the poor, the distributional benefits of some aid <u>can</u> be improved without much efficiency loss, by channelling it down to lower levels, for example through NGO's and targetted credit programs.

12.   While there is no easy solution to poverty in Nepal, the potential contribution of public policy is large. As an illustration - if the Government is able to achieve the best reasonable expected results from an effective family planning program and in sustaining economic growth, then per capita GDP could rise to about $270 equivalent per capita by 2010 - probably holding the number of poor to below 5 million. On the other hand, if population growth continues unabated, and GDP growth is no better than the average achieved over the last twenty years, then per capita incomes would stagnate at around $180 per annum - probably increasing the number of poor to over 20 million by 2010. In short, the cost of not getting public policy right - especially with respect to curbing population growth - is probably in the order of 15 million additional absolute poor.

13.   The <u>priority elements of a poverty alleviation strategy</u> in Nepal need to consist of:

(i) an effective program to curb population growth - through the promotion and support of temporary methods of contraception;

(ii) an agricultural program that includes: small farmer irrigation (and measures to get all irrigation working in the terai), input supply deregulation, development of a technical package free of purchased inputs for inaccessible hill farmers;

horticultural support for accessible small farms in the hills; and design of an outreach cum extension program that takes account of the constraints facing poor farmers;

(iii) a program of increased rural access in the terai and selected areas of the hills;

(iv) intensification of basic education, including: (a) a national literacy campaign; (b) revamping of curriculum and examination systems to improve relevance and quality; (c) measures to lift constraints on attendance by girls and by the poor; and (d) agreement on commitment of adequate financing for expansion of the primary school system to provide almost universal coverage; and,

(v) in addition, there is scope for a package of low-cost measures to help the large numbers of those who will remain absolutely poor for the foreseeable future - through selected off-farm income generating activities and improvements in health, nutrition, and access to food (A program of such poverty alleviation measures is described in paragraph 16, below).

14. Conversely, there are a number of areas in which the Government should probably exercise caution, and a further sub-set of areas in which action is justified, but where further analysis is needed first. The following are areas in which, on the basis of the analysis, it appears that public interventions would not be a cost-effective means of helping the poor:

- large-scale infusions of targetted credit;

- price controls or commodity subsidies; and

- excessive intervention in the industrial sector.

Those areas in which further analysis is needed to formulate effective programs include:

- partitioning the hills for poverty alleviation purposes into areas for varying degrees of support, depending on the efficiency of providing access at reasonable cost;

- cost-effective measures to improve access to food in remote areas;

- analytical work to develop a strategy for managing the transition of population from the hills to the terai and to urban areas;

- an assessment of the scope for effective measures to promote income-earning opportunities in the informal sector;

- investigation of the scope for, and implications of, reforms to the land tenure system, to improve both efficiency and equity.

- improvements in the design and implementation of income-generation and skills training projects.

15.   Pervasive poverty in Nepal stems from four factors: (i) a limited resource base; (ii) a physical location between two large countries - both of which are also poor; (iii) rapid population growth, and (iv) poor economic performance - with GDP growth averaging under 3% p.a. over the last 25 years. Little can be done about the first two factors, however effective policies can reduce population growth and accelerate economic expansion. Unfortunately Government initiatives in these areas to date have been partial and ineffective. The Government now has an opportunity to play a more aggressive role - by pursuing a very active population program, and laying the basis for more rapid growth, through, <u>inter alia</u>: widespread public service reforms, accelerating education and skills training, further liberalization of trade and industrial restrictions, a more aggressive agricultural growth strategy, and reform of the financial sector. Additional measures are required to ensure that the poor specifically benefit, and these are discussed in the following sections.

16.   The report concludes that even with the best expected performance in economic growth and family planning, there are likely to remain some 5-10 million absolute poor over the next twenty years. It therefore proposes, in addition to the broad priorities for creating equitable growth outlined above, a set of measures for Government and donor support specifically to assist those who will inevitably remain poor, consisting of the six elements outlined below:

(1)   <u>Expansion of Selected Transfer Programs</u>. These are large-scale rural employment programs which, in the absence of sufficient growth in the medium-term, can provide cost-effective transfers while at the same time creating productive rural assets. Two such programs exist in Nepal - Food for Work, and the Special Public Works Program - with some strengthening they could be amenable to substantial expansion. The next step would be to determine how much they could expand, their financial requirements, to identify measures to improve targetting and efficiency, and agree on a financing package for them.

(2)   <u>Incremental Improvements to Existing Income-Generating Programs</u>. There are a number of smaller programs which have had some success in raising incomes of the poor (including the Small Farmers Development Program, Production Credit for Rural Women, and localized NGO income-generating projects). While these can never have sufficiently wide coverage to make a major reduction in the incidence of poverty, they should be expanded as staffing constraints allow. There is also room for incremental improvements to their operations, including more intensive training of staff and better sub-project evaluation.

(3)   <u>Food and Food Aid Programs</u>. Enough areas have sufficiently large food deficits, which cannot be solved by out-migration or agricultural productivity increases in the near term, that food aid and food distribution programs will be an important continuing part of any poverty alleviation strategy. The objective now should be to identify the most cost-effective combination of interventions (eg. relying on increased local

production, improved feeding practices, etc. whenever possible), and to rationalize existing programs, which often do not reach the intended beneficiaries. This would include:

- agreement on the appropriate level and form of food aid, its financing, and how best to distribute it;

- reform of the National Food Corporation's program - eliminating mis-targetted subsidies, identifying which distribution measures reach those suffering food deficits, and how they should be expanded and financed; and,

- identification of a program of effective interventions - including evaluating the relative roles of food distribution, agricultural productivity measures, food storage, vulnerable group feeding, and promotion of effective practices (e.g., weaning foods, feeding during pregnancy, etc.); followed by preparation of specific projects and agreement on financing for them.

(4) Strengthening Non-Income Welfare Measures. These are interventions which could substantially improve the welfare of the poor even in the absence of increased incomes. They include:

- a national hygiene campaign - which would combine intensive hygiene-awareness education with the provision of rural water supplies;

- a range of particularly cost-effective health measures - (immunization, iodine supplementation, and oral rehydration salts) - these are existing programs which could be expanded at modest cost without awaiting major improvements in the (institutionally weak) health service; and,

- (possibly) targetted nutrition and/or feeding programs - these can have a dramatic effect on reducing the long-term effects of malnutrition at affordable cost. Such programs usually rely on an administrative capacity which may not be present in Nepal; however, it is worth exploring the possibilities.

(5) A Package of Policy Reforms.

Tenancy Reform - a redistributive land reform would have some impact in the terai, but may not be feasible. However, the existing tenancy system provides many disincentives to maximizing output. It is possible to design a reform which at a minimum would regularize land tenure and remove barriers to efficient use of land. This, in the terai, could have a large impact on agricultural productivity. However not enough is known about the distribution and types of tenancy to design a workable reform immediately - some analytical work is required first.

Labor Contracting Arrangements - current arrangements are exploitative beyond what would be expected on the basis of low average wage rates alone; some improvements would be possible by introducing revised procedures for public works (eg. the use of small local contractors) and increased monitoring and supervision.

Decentralization and Civil Service Reform - effective implementation of the decentralization process can potentially help the poor by improving the delivery of social and development services to peripheral areas (by devolving staff and resources to them). Wider civil service reform is also needed, but it will probably be a slow process. At a minimum administrative improvements can be made which will increase the effectiveness of public services (by for example strengthening supervision and incentives for actively providing outreach services).

NGO Regulation - the current system has constrained NGO activity and probably added to their costs. Government should explore the scope for relaxing the regulatory framework to facilitate NGO operations - along with measures to improve the technical support provided to them (for instance, to improve project selection and design).

(6) Analytical Work On Priority Areas. This includes assembling data on incomes, wages and employment, and further analysis needed to design policy and program reforms.

I. INTRODUCTION

A. Objectives and Scope

1.1   This study is intended to deepen our understanding of the nature of poverty in Nepal; of its causes, and of the constraints which prevent the poor from improving their conditions. It investigates the effect of development policies and strategies on personal incomes, and seeks to identify the most promising areas for raising incomes of the poor. The objectives are to propose the outlines of a long-term country strategy to reduce poverty, as well as to recommend specific measures for Government and donor support.

1.2   The study came about as the result of a number of perceptions:

> Firstly, that extreme poverty is a widespread problem in Nepal, but that little analytical work had been done on income levels and trends, or the dynamics at work among the poor.

> Secondly, that there exists a desire within the donor community, and within the World Bank, to do more to support poverty alleviation efforts in Nepal, but that it has been difficult to identify which measures might be amenable to large-scale financial support.

> Thirdly, Nepal is a very poor country in which the Bank and the donor community have made substantial investments (presumably with the objective of raising living standards), but with little understanding of how our programs affect the incomes of the poor. The study seeks to better inform overall country assistance by investigating the links between incomes of the poor and various agricultural, industrial, and employment activities.

> Fourthly, the Government is already putting substantial resources into programs notionally justified on poverty-alleviation grounds - the analysis looks at ways these resources could be used most cost-effectively.

> Finally, in Nepal there have been many good village-level and sectoral studies, each of which tells an important part of the poverty story, but the isolated findings have never been pulled together.

1.3   The study has sought to assimilate the findings in various sectors, and build them into a wider analytical framework. The objectives are:

    (i)   to outline the bounds of what might be possible (and expected) in the evolution of incomes over the next 20 years, and to evaluate the relative contributions of agriculture, industry and commercial development, and social services to reducing poverty;

(ii) to suggest sectoral interventions, or changes in approach, which may have a greater impact on incomes, especially among the poor; and,

(iii) to propose a menu of cost-effective poverty alleviation projects for possible support.

1.4   The report is organized as follows: Chapter II analyzes the levels, distribution and sources of income. It presents for the first time in quantitative terms a profile of poverty in Nepal, characteristics of the poor and non-poor, and the conditions under which they live. This analysis is based on special tabulations of the Nepal Rastra Bank's household survey data undertaken as part of this study. Chapter III explores the causes of poverty - both those related to aggregate output and growth, and the more subtle mechanics of land tenure, employment, and social structures. Chapter IV looks at linkages between the productive sectors (agriculture, industry, etc.) and incomes, and the role of non-formal employment. Chapter V examines the social dimensions of poverty: food security, education, and health and population issues. Chapter VI briefly reviews His Majesty's Government's (HMG's) general policies, and summarizes our assessment of poverty-related programs based on a more detailed earlier version of this report.1/ The final two chapters suggest implications for country strategy and donor support.

1.5   Of necessity the report takes a longer-term perspective (20-30 years) than is usual in World Bank documents. It also covers a wide range of sectors. In not all of these were we equipped to bring our conclusions to closure; in some we could only highlight the issues involved and point to possible solutions, or areas for further work.

1.6   Other reports on poverty have tended to focus on social services and their delivery. These issues have been explored at length in an earlier Bank study.2/ We have concentrated instead on the objective of raising personal incomes and consumption.3/ While this may provide a somewhat one-dimensional picture, it was necessary to limit the scope of the work to something manageable, and furthermore raising personal incomes (broadly defined) is, after all, the single best measure of reducing poverty.

___

1/ The World Bank; <u>Nepal - Relieving Poverty in a Resource-Scarce Economy</u>, 1990.

2/ The World Bank; <u>Nepal - Social Sector Strategy Review</u>, 1989.

3/ Incomes are taken throughout to include production of food for own consumption - which constitutes by far the largest share of personal income in Nepal.

B. **The Nature of the Problem - Poverty in Nepal in a Comparative Context**

1.7   Average personal incomes are about Rs. 3,340 per capita (1984/85, expressed in 1988/89 prices) - equivalent to US$130 annually. Approximately 40% of the population live in absolute poverty, defined as having less than the income required to consume a minimum bundle of calories on a daily basis. There are also chronic and seasonal food deficits, which probably affect half of the population. Large scale starvation is avoided because of the absence of catastrophic droughts, although slow physical deterioration through malnutrition is widespread.

1.8   In comparative terms, Nepal is one of the world's poorest countries. It ranks 115th in per capita GNP out of 120 countries listed in the 1989 World Development Report (WDR). With respect to life expectancy, Nepal ranks 103rd out of 118 reporting countries. Food availability, measured in calories per capita, is worse only in Bangladesh, Haiti, and a handful of African countries. Table 1.1 presents key development indicators for Nepal, relative to some of the world's poorest countries, and relative to neighboring states in Asia.

Table 1.1: Indicators of Relative Underdevelopment - Selected Countries 1/

	Life Expectancy (Years)	Per Capita GNP (US$)	Infant Mortality (per 000)	Population Growth Rate	Calorie Availability (Calories per capita per day)
Ethiopia	47	130	154	3.1%	1749
Mali	47	210	169	3.0%	2074
Nepal	51	160	128	2.5%	2052
Bangladesh	51	160	119	2.4%	1927
Zaire	52	150	98	3.1%	2163
India	58	300	99	1.8%	2238
Thailand	64	850	39	1.5%	2331
Sri Lanka	70	400	33	1.1%	2401

1/ According to the 1989 World Development Report (WDR). For consistency of inter-country comparisons all data are drawn from the 1989 WDR; these may not be consistent with estimates for Nepal cited elsewhere in the text, which were drawn from more recent surveys and reports.

1.9   Nepal is one of a constellation of countries characterized by rapid population increases, low or negative per capita GDP growth, and a slow transition out of a subsistence agricultural economy. It exhibits many of the characteristics of similar sub-Saharan African economies, including a limited productive land base, a land-locked location, and a very low level of exports. Table 1.2 illustrates some of the structural features of Nepal, relative to other countries.

Table 1.2: Structural Economic Indicators

	Per Capita GNP	Per Capita GNP Growth Rate 1965-87	Share of Agriculture in GDP	Per Capita Merchandise Exports	Per Capita Commercial Energy Consumption (kg. oil equivalent)
Very Poor Countries					
Ethiopia	$130	0.1%	42%	$9.00	21 kg.
Mali	$210	0.1%	54%	$27.70	24 kg.
Zaire	$150	-2.4%	32%	$48.90	73 kg.
Nepal	$160	0.5%	57%	$8.60	23 kg.
Neighboring Asian Countries					
Bangladesh	$160	0.3%	45%	$10.10	47 kg.
India	$300	1.8%	30%	$15.70	208 kg.
Thailand	$850	3.9%	16%	$217.50	330 kg.
Sri Lanka	$400	3.0%	27%	$89.90	160 kg.

All data are for 1987 unless otherwise indicated.

1.10   Nepal's share of agriculture in GDP, one of the best summary indicators of the level of development and structural change, is, at 57%, higher than all but three of the very poorest countries (Uganda, Somalia and Tanzania). Similarly, the share of the labor force dependent upon agriculture (93%) is the highest of any country listed in the 1989 WDR. Nepal is also one of the two lowest consumers per capita of commercial energy - a sensitive indicator of industrial and urban development. Nepalese consume 23 kg. of oil equivalent annually compared with 297 kg. for low income countries as a whole.

1.11   There is a very high cost in human suffering and wasted potential associated with Nepal's degree of poverty. For example, about 6 1/2% of the population is estimated to suffer some degree of mental retardation, due mainly to malnutrition and iodine deficiency. Again, by the Indian Academy of Pediatrics standard, about 29% of the population exhibit 2nd or 3rd degree malnutrition. Associated with nutritional deficiency and diseases, there are estimated to be about half a million cases of blindness in Nepal (out of a population of about 18 million in 1988), a rate far in excess of international standards.

II. THE STRUCTURE OF INCOMES AND POVERTY IN NEPAL

A. Levels and Distribution of Income

2.1   Average per capita incomes are estimated at about Rs. 260 per month in 1988/89 prices (US$122 p.a.) in rural areas; and about Rs. 426 (US$200 p.a.) in urban areas. In rural areas, the mean income in the hills is about 20% lower than in the terai 1/ ; in urban areas, reflecting in particular the relative wealth of Kathmandu, the order is reversed, with average incomes in the hills about one quarter higher than in the terai.

2.2   Within Nepal, the best source of data relating to incomes is the Multi-Purpose Household Budget Survey (MPHBS) conducted in 1984-85 by the Nepal Rastra Bank.2/ The analysis in this chapter is based on a re-processing of the MPHBS data undertaken by the World Bank specifically for this study to identify the characteristics of poor and non-poor households. Average per capita monthly income levels by decile are presented in Table 2.1. These data are expressed in terai-equivalent prices, so that they can

Table 2.1: Average Per Capita Monthly Income by Income Deciles
(1984/85 expressed in 1988/89 Rs per month)

	RURAL		URBAN	
Decile	Terai	Hills	Terai	Hills
1	123	85	162	171
2	164	116	203	229
3	188	140	235	281
4	207	163	262	334
5	232	188	294	395
6	262	217	343	453
7	295	253	396	543
8	348	299	493	644
9	435	368	675	830
10	719	564	1196	1394
All Families	292	231	378	475

Source: Mission estimates derived from MPHBS.

1/ The terai is a narrow lowland belt along the Indian border containing about half of Nepal's population. Incomes here are adjusted for differences in purchasing power; consumer goods flow mainly from the terai to hills and transport costs are very high. Mission estimates of the price differentials, based upon a basket of eight staple foods, suggest that prices are about 32% higher in the hills than terai.

2/ The survey is described in Annex II.2. While the MPHBS is an exceptionally well-designed and well conducted survey, it suffers from the inability of all such surveys to capture fully the exceptionally rich and the exceptionally poor. For this reason, the actual income distribution is more skewed than that revealed in the tabulations.

be compared across hills and terai. Incomes are uniformly low except in the top decile - and are even then only about two and a half times the average.

2.3   Like other very poor countries, Nepal's predominantly rural economy sustains a relatively even distribution of income, corresponding to a stage of development before the income and wealth-concentrating process of urban-based development has had much impact. From Table 2.2 it can be seen that Nepal has a degree of income concentration somewhat more favorable to the poor than Sri Lanka, somewhat less so than, but comparable with, Sweden, and markedly more egalitarian than the distribution of income in Brazil.

Table 2.2: Distribution of Total Household Income in Nepal and Selected Countries
(Percentage of Share of Income)

Household Group	Nepal	Sri Lanka	Brazil	Sweden
Bottom 40%	18%	16%	7%	21%
Middle 50% /a	54%	49%	42%	51%
Top 10%	28%	35%	51%	28%

a/   That is, lying between the 41st and 89th percentiles.

Source: Nepal, MPHBS. Other countries, World Development Report, 1989.

2.4   These comparative data suggest that the scope for improving the income position of the poor in Nepal by redistribution from the upper income groups is limited. For example, in Brazil, transferring 5% of the income of the upper decile to the lowest 40% would involve an income increase for the latter of about 36%, an amount which would have a significant impact in reducing the level of poverty. In Nepal, by contrast, transferring 5% of income from the richest 10% to the bottom 40% would increase the average incomes of the latter group by less than 8%, a welcome improvement no doubt, but not one which would produce a major impact upon poverty levels. At the same time it should be noted that the position of the poor in Nepal could be worsened greatly by a deterioration of the distribution of income. As discussed later in this report, there are reasons to believe that the mechanisms exist for such a worsening, which can only be countered by sustained off-setting policy interventions.

2.5   The distribution of per capita household income is presented in Table 2.3. As expected, it is less skewed than that of total household income (Table 2.2). (This is because household size increases consistently with total income, from less than 4 persons per household in the lowest income class to over 13 in the highest). The scope for poverty alleviation by redistributive measures alone is, therefore, even less than that suggested by the cross-country data contained in the previous table.3/

Table 2.3: Per Capita Household Income Distribution in Nepal
(percentage share of income)

Income Group	All Nepal	Rural Terai	Rural Hills	Rural Mountains	Urban Terai	Urban Hills
Bottom 40%	23%	24%	23%	33%	27%	24%
Middle 50%	54%	53%	56%	54%	52%	56%
Top 10%	23%	23%	21%	13%	21%	20%

Source: Multi-Purpose Household Budget Survey.

2.6 The distribution of income is essentially the same in all regions except the mountains where it is remarkably flat, with poverty being shared by almost the entire population. However, the proportion of Nepal's population which lives in the mountains is very small, and mountain people are included within the hills for much of the subsequent discussion.

The Incidence of Poverty

2.7 Table 2.4 shows the proportion of the population below the poverty line under three different sets of assumptions. The first is the poverty line defined by the National Planning Commission (NPC) on the basis of an income needed to supply minimum caloric requirements. In 1988/89 Rupees, this translates into Rs. 210 per person per month in the hills and Rs. 197 in the terai, or the equivalent of US$8.24 and $7.73 respectively per person per month. With this poverty line, the data suggest that about 40% of the population is in absolute poverty.4/ That this is a very conservative definition of poverty can be seen by considering the implications of a poverty line equivalent to US$150 per capita per annum.5/ The proportion of the population with incomes below even this modest level is about 70% nation-wide, and almost 80% in the hills. This calculation suggests that calorie-based poverty lines may not be appropriate for Nepal, a matter which is given additional support by the fact that the usual relationship between food intake and measured under-nutrition in Nepal is weakened by the widespread prevalence of intestinal disorders (see Chapter 5).

3/ Note, however, the reservation contained in the earlier footnote: the data, like those from other such surveys, under-represent households at the highest income levels.

4/ The incidence data are approximations only. Note that the imputed rental value of owned housing was not included in total expenditures in these calculations.

5/ An accepted international definition of absolute poverty. (1988-89 rupee equivalent used; average exchange rate Rs. 25.5/US$1.00).

2.8   A third possible definition is one suggested by Lipton.6/ He has defined the poor as those whose food expenditures absorb 70% or more of total expenditures. This criterion underlies the third poverty line in Table 2.4, in which the Nepal-wide incidence of poverty is about 66%, similar to the incidence implied by a poverty line of US$150 a year.

Table 2.4: Incidence of Poverty Under Different Poverty Lines
(Percentage of Population Below Poverty Line)

Poverty Line	Terai Rural	Terai Urban	Terai Total	Hills Rural	Hills Urban	Hills Total	All Nepal Rural	All Nepal Urban	All Nepal Total
NPC	29%	17%	28%	55%	13%	52%	42%	15%	40%
US$150	69%	51%	68%	78%	32%	75%	74%	42%	71%
Lipton	70%	50%	68%	65%	52%	64%	68%	51%	66%

2.9   The sensitivity of the number of poor to the choice of poverty line results from the flat distribution of income around a very low average. Thus moving the line up or down only slightly results in large changes in the share of the population above or below it. Within the limitations of the data it is thus not very meaningful to talk of the precise number of poor. The best one can conclude is that the incidence of absolute poverty is very high - in the neighborhood of 50-60%. In subsequent discussion we will use the (conservative) NPC poverty line.7/ By this measure about 8 million Nepalese live in absolute poverty - out of an estimated total population of 19.2 million.

2.10   While in the following sections reference is made to the "poor" and the "non-poor", it is important to recognize that by any reasonable international standard everyone in Nepal is poor, except for a few professionals and businessmen, and perhaps some large farmers. The average income in the second decile (i.e. the second richest 10% of households) for instance is only about Rs. 500 per capita per month (US$20 per month) - in most countries these families would be among the very poorest.

2.11   It is often useful to distinguish between the poor and a subgroup of "ultra-poor" - at whom one may want to target assistance or development. In Nepal such a distinction is not particularly useful - both because there are so many very poor (eg. about half the population), and because most of the poor exhibit characteristics usually associated with the ultra poor (eg. high proportions of income spent on basic cereals). However, while they are frequently hungry, they are not ultra-poor in the longer-term sense that

---

6/ M. Lipton: Poverty, Undernutrition, and Hunger; World Bank Staff Working Paper No. 597, 1983.

7/ In purchasing power parity terms this is estimated to be equivalent to about US$328 per capita p.a. - in the middle of the range of poverty lines cited in the 1990 World Development Report ($270-$375).

they are an under-class caught in a chronic trap, out of which they could not raise themselves with access to technology and sufficient improvements in productivity. There is, however, a growing group of such ultra-poor emerging among the urban population, and among the landless in the terai.

B. Composition of Income

2.12   Figure 1 illustrates the overall composition of income emerging from the MPHBS. Of note are the overwhelming importance of subsistence activities and of agriculture, and the relative unimportance of wage incomes.

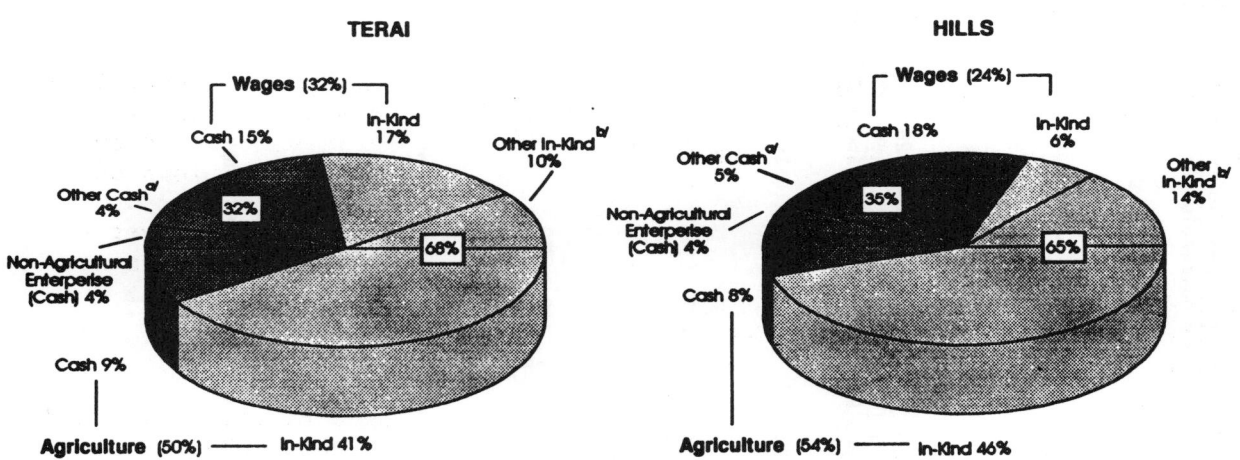

Figure 1.
Composition of Income Among the Rural Poor

a/ Mostly remittances, pensions, etc.
b/ Mostly rent-free dwellings and cottage production for own consumption.

2.13   Table 2.5 shows the composition of income for different groups. The following particular characteristics emerge:

- among the rural poor, only 35% of incomes are in cash (32% in the terai) - coming mostly from wages;

- the poor get a larger share of their incomes from wages and salaries than do the non-poor (who make <u>much</u> more from agriculture, especially in the terai);

- even then wages and salaries account for only a quarter of income among the rural poor in the hills (a bit less than a third in the terai);

Table 2.5: Composition of Household Income - 1984-85
(1984/85 Rs. per household per month)

	RURAL					URBAN			
	Terai		Hills		Terai		Hills	
	Poor	Non-Poor	Poor	Non-Poor	Poor	Non-Poor	Poor	Non-Poor
Agriculture								
Cash	64 (9%)	398 (27%)	56 (8%)	262 (18%)	26 (4%)	90 (6%)	52 (6%)	83 (4%)
Kind	285 (41%)	684 (46%)	306 (46%)	628 (43%)	160 (23%)	271 (24%)	197 (24%)	220 (12%)
Subtotal	349 (50%)	1082 (73%)	363 (54%)	890 (61%)	186 (26%)	361 (26%)	248 (31%)	303 (16%)
Non-Agricultural /a								
Enterprises	29 (4%)	80 (5%)	28 (4%)	121 (8%)	106 (15%)	377 (27%)	77 (10%)	472 (25%)
Wages and Salaries								
Cash	108 (15%)	107 (7%)	121 (18%)	209 (14%)	300 (43%)	417 (30%)	350 (43%)	721 (38%)
Kind	120 (17%)	52 (14%)	39 (6%)	19 (1%)	34 (5%)	21 (2%)	12 (1%)	19 (1%)
Subtotal	228 (32%)	160 (11%)	159 (24%)	228 (16%)	334 (48%)	438 (32%)	362 (44%)	740 (39%)
Other Cash Income	24 (4%)	63 (4%)	30 (5%)	96 (7%)	30 (4%)	143 (10%)	46 (6%)	339 (18%)
Other Income In-kind /b	71 (10%)	108 (7%)	91 (14%)	119 (8%)	47 (7%)	59 (4%)	83 (10%)	66 (3%)
Total /c	702 (100%)	1493 (100%)	671 (100%)	1452 (100%)	704 (100%)	1379 (100%)	815 (100%)	1919 (100%)
Of Which:								
In Cash:	224 (32%)	646 (43%)	236 (35%)	679 (47%)	464 (65%)	1023 (74%)	523 (64%)	1603 (84%)
In Kind:	478 (68%)	845 (57%)	437 (65%)	773 (53%)	245 (35%)	357 (26%)	292 (36%)	315 (16%)

a/ Almost all of which is in cash.

b/ Home production for own consumption plus rental value of rent-free dwellings.

c/ Excludes rental value of own home.

- even in urban areas wages account for less than half of the incomes of the poor;

- a significant portion of wage incomes of the poor is in kind rather than cash (over half in the terai);

- cash incomes from agriculture (representing sales of foodgrains, cash crops or livestock products) are insignificant among the poor - less than US$3.50 equivalent per household per month (8-9% of income); and,

- off-farm enterprises (eg. small businesses, etc.) are also insignificant - contributing less than 5% of household incomes among the rural poor ($1.50/household/month) - and even in urban areas agriculture remains three times more important than off-farm enterprises as a source of income.

2.14   Localized surveys tell much the same story as the MPHBS, but provide some additional insight into sources of income (see Table 2.6).

Table 2.6: Some Survey Estimates of Income Composition
(Share of household income)

8 Villages - Acharya and Bennett	Average	Rural Terai	Rural Hills
Subsistence: Agriculture	48%	61%	45%
Kitchen Gardening	2%	4%	2%
Livestock	8%	6%	10%
Hunting and Gathering	5%	2%	6%
Food Processing	15%	17%	12%
Household Manufacture	1%	1%	2%
Subtotal Subsistence	81%	92%	79%
Non-Subsistence: Trading	7%	1%	3%
Wages and Salaries	12%	7%	18%

Two Hill Districts - Rasuwa Nuwakot IRDP

	Low Income Households	High Income Households
Agriculture	69%	89%
Of Which: Livestock	(4%)	(34%)
Cottage Industry	-	1%
Trade	3%	2%
Wages	15%	2%
Portering	9%	2%
Remittances, Pensions, etc.	4%	4%
Subtotal - Off-Farm Income	31%	11%

2.15   In the sample surveys livestock contribute about 5-10% of income (except in the case of high income families in the hills sample - where they contributed a third). In no cases do wages and salaries account for more than about a quarter of household income. Cottage industries and trade are insignificant as sources of income, while portering was important only in one of the hill districts studied (where it accounted for about a third of off-farm income among the poor).

2.16   The relative importance of off-farm income among the poor (about 40%), does not reflect their active participation in a robust informal sector. Instead it seems to be generated mostly from a range of distress activities and very low paying employment. The fact that it constitutes a relatively large proportion of income is a function of the very low _total_ incomes among the poor, and the very low levels of own-farm production, rather than an indication that the poor are doing well out of off-farm employment.

2.17   While the size of landholdings is a major determinant of rural incomes, the influence is not straightforward (see Table 2.7). Firstly, family size increases with the size of farm, so that while per capita income increases with farm size, it does so less than proportionately. Secondly, returns to family labour are higher on large farms, due to diminishing returns to labour, and a more favorable land/person ratio. Thirdly, while small landholders are _much_ more dependent on off-farm income, they earn substantially less per day worked off-farm - less than half as much as those from larger farms (Table 2.8).

Table 2.7: The Effect of Farm Size on Incomes
(1984/85 Rs.)

	Large	Medium	Small	Marginal	Non-Cultivator
Terai					
Farm Size (ha.)	>5.4	2.7-5.4	1.0-2.7	<1.0	-
Percent of Households	3%	9%	23%	38%	27%
Average Household Size	12.2	9.0	7.0	5.5	4.7
Total Monthly Income	3380	1822	1210	787	633
Per Capita Monthly Income	277	202	174	143	136
Hills					
Farm Size (ha.)	>1.05	0.5-1.05	0.2-0.5	<0.2	-
Percent of Households	16%	24%	50%	11%	-
Average Household Size	7.9	6.1	5.0	4.3	3.8
Total Monthly Income	1284	841	635	462	559
Per Capita Monthly Income	162	139	128	108	150

Table 2.8: Returns to Family Labor Per Day Worked By Farm Size
(Rs./day)

	Large	Medium	Small	Marginal	Non-Cultivator
Terai					
Agriculture on Farm	58	42	36	34	-
Other	103	62	25	15	14
Hills					
Agriculture on Farm	40	34	35	31	-
Other	56	31	26	20	20
Percent of Family Labor Time Spent in Off-Farm Work /a					
Terai	5%	8%	22%	65%	100%
Hills	13%	24%	39%	67%	100%

a/ Share of employed days

2.18 This decline in off-farm earning rates, combined with the increased dependence upon off-farm earnings on the part of small farmers, greatly reinforces the poverty-inducing impact of lower earnings from work on the farm itself. One likely explanation of the lower earning rates lies in the far lower educational attainments of family members on small farms. Breaking this relationship must be an important element of poverty alleviation efforts in Nepal.

C. Profile of the Poor

2.19 The poor are concentrated overwhelmingly in rural areas (95%) and more in the hills than in the terai (because food costs are significantly higher in the hills; nominal incomes unadjusted for price differentials are in fact lower in the terai).[8] The principal characteristics of the poor (and the non-poor) are shown in Table 2.9.

---

[8] Here, and in subsequent discussion, the poor are defined in accordance with the NPC poverty line as having per capita household incomes below Rs. 197/month in the terai and Rs. 210/month in the hills (1988/89 prices). The profiles draw on special tabulations of the MPHBS data done for this study - many of which are presented in Annex II). While this is a very conservative poverty line which may in fact select only a sub-group of the poor, it does not materially change the profile that emerges.

Table 2.9: Nepal - Poverty Profile

	RURAL NEPAL		URBAN NEPAL	
	Terai	Hills	Terai	Hills
Estimated No. of Poor	2.9 m.	4.6 m.	0.3 m.	0.3 m.
Economic Characteristics				
Average per capita Income				
Rs./day	3.4 (7.2)	3.9 (8.8)	3.4 (8.2)	4.6 (12.3)
US$ p.a.	$69 ($148)	$81 ($181)	$69 ($168)	$95 ($253)
Operating landholding (ha.)	1.1 (3.1)	0.3 (0.5)	-	-
Proportion of Family Labour Days Worked Off-Farm	31% (14%)	17% (10%)	36% (20%)	39% (21%)
Occupation /a				
Agriculture Self-Employed	54% (67%)	74% (74%)	42% (31%)	46% (29%)
Agricultural Labourer	28% (12%)	7% (2%)	4% (3%)	3% (-)
General Labourer	8% (3%)	11% (6%)	22% (9%)	30% (7%)
Other /b	10% (18%)	8% (18%)	32% (43%)	21% (64%)
Demographic Characteristics				
Household Size	7.1 (6.8)	6.1 (5.4)	6.9 (5.5)	6.3 (5.1)
Number of Children	3.5 (2.7)	2.9 (2.1)	3.5 (2.2)	3.2 (2.0)
Dependency Ratio /c	2.3 (2.1)	1.9 (1.7)	2.7 (2.7)	2.7 (2.7)
% of Female-Headed Households	5% (4.5%)	12% (11.5%)	5% (6%)	15% (15%)
Literacy Rate	22% (40%)	39% (51%)	35% (59%)	49% (72%)

a/ Proportion of economically active household members by main occupation.
b/ Sales, Services, Production, Clerical and Professional.
c/ Number of household members per employed member.

NOTE: Corresponding figures for the non-poor are presented in parentheses.

Source: MPHBS special tabulations, incomes in 1984/85 Rs.

2.20 *Occupations*. The poor, like everyone else in Nepal, are engaged mostly in agriculture on their own (or rented) land. The only notable exceptions are in the terai - where 28% of the economically active poor are employed by others as agricultural wage labourers, and in urban areas where about 25% are engaged as general labourers, and another 25% in services or other miscellaneous activities. Even in urban areas the largest proportion of the poor are engaged in agriculture as their principal pursuit (more so than among non-poor urban dwellers). Only about 5% of the active poor are employed in production or manufacturing jobs of any kind - including rural cottage industries.

2.21   Most people, and especially the poor, are employed only sporadically, even in their main occupation. Although the poor on average work twice as much time off-farm as the non-poor, they are still working only 21% of available labour days off-farm (higher in the terai, lower in the hills). There is correspondingly high underemployment - reportedly in the range of 35-45% of available labour days, with little distinction between the poor and non-poor (although this represents some under-counting of subsistence activities - see discussion of labour force issues in Chapter 3).

2.22   In rural areas, the main income-producing asset is of course land, and access to land largely determines income levels. The area of land operated by a poor household is about 60 percent less than that operated by a non-poor one in the terai, and about 40 percent less in the hills. In both regions, the proportion of the land which is irrigated is somewhat less for the poor than for the non-poor household, although the differential access to irrigated land reported in the MPHBS is rather less than is commonly supposed.9/

2.23   <u>Household Composition</u>. The average size of poor households is slightly greater than that of non-poor ones. Consistently, also, there are more children in poor households; and a tendency for more children per household in the terai than in the hills. The relative concentration of children in poor families means that the children of Nepal are proportionately more exposed than are adults to all the disadvantages of poverty - including limited access to food, education, health services and sanitation. At the same time, they are more vulnerable than are adults to the consequences of poverty.

2.24   The proportion of female-headed households in the hills is quite high (about 11-15%.) due to temporary and long-term out-migration. These proportions seem to differ little by income level or degree of urbanization. Reflecting the same basic phenomenon, the ratio of working-age adult males to females is much lower for poor families than for non-poor ones, although it is not clear if this is cause or effect (see para. 3.90).

2.25   Finally, the dependency ratio (or the ratio of total household members to earning household members) shows some unusual patterns. Usually, a higher dependency ratio (many dependents per earner) characterizes poor families. This pattern shows up weakly in Nepal, reflecting mainly the larger number of children in poor families discussed above. However, the least poor category (the urban areas) exhibits higher (rather than the lower) average dependency ratios. The explanation lies in the fact that in the poorer areas, more of the family, and particularly children and women, are pressed into becoming earners than is the case in less-poor areas.

9/   See Annex II.6; this may be due in part to the reporting of all <u>khet</u> land as irrigated - where as in fact much is only monsoon-irrigated.

D. Conditions of the Poor

2.26 This section attempts to convey a sense of the conditions under which the poor live in Nepal. It draws largely on the MPHBS and can therefore be regarded as providing profiles which are broadly representative of large groups in Nepal. They are in this regard to be distinguished from accounts of living conditions among particular communities in particular localities - accounts which are often strikingly vivid, but which by their nature may lack general applicability.

Expenditure and Consumption

2.27 The poor are consuming a minimal bundle of goods, the cost of which is often not covered by their available incomes. Food (much of it self-produced or received as wages in kind) accounts for 70-80% of the income of the poor, leaving very little for expenditure on other essentials, let alone to finance diversification. The following table summarizes key aspects of discretionary income.

Conditions of the Poor - Summary Table - Expenditure
(1984/85 Rs. per capita per month)

	RURAL NEPAL		URBAN NEPAL	
	Terai	Hills	Terai	Hills
Monthly Consumption				
Expenditure	105 (185)	121 (234)	110 (202)	155 (323)
US$ Equivalent	$5.90($10.40)	$6.80($13.15)	$6.20($11.35)	$8.70($18.15)
Discretionary				
Income a/	23 (110)	38 (128)	22 (147)	30 (200)
Discretionary Cash				
Income b/	14 (77)	23 (98)	22 (123)	23 (189)
US$ Equivalent	$0.79($4.32)	1.29($5.50)	$1.24($6.90)	$2.19($10.60)

NOTES: Corresponding figures for non-poor shown in parentheses. Not adjusted for price differentials - hills purchasing power about 30% lower than terai.

a/ After food; approximately 40% lower after fuel and water purchases.
b/ After food purchases.

2.28 The table provides two somewhat different indications of upper bounds upon the capacity of the poor to pay for public services under the "user pays" principle. The first is the level of discretionary income, or the income remaining after the food (but not fuel or other) needs of the family are met. The per capita discretionary income among poor households ranges from about Rs. 20 to Rs. 40 monthly, or about $1.25 to $2.50 - and is about 40% lower after fuel and water costs. The second is the availability of cash incomes. Cash incomes of poor families vary between about Rs. 30 and Rs. 45 per capita monthly in rural areas, and somewhat higher in urban areas. After essential food purchases (which even then do not bring them up to minimum caloric intake levels) available cash incomes range for Rs. 14 to Rs. 23 per capita monthly (equivalent to US$0.80 to US$1.30). The capacity of rural poor households to contribute significantly to the costs of public services is thus severely limited, as is their ability to purchase cash inputs, or finance investments which could ultimately raise their earning power (for instance in land improvement or education). As an example, the cost of applying fertilizer to only _half_ of the land operated by a typical poor terai household would be in the neighborhood of Rs. 500 - compared with discretionary cash incomes of only about Rs. 100 per month, a large share of which is likely to be required for cooking fuel purchases. This absolute lack of discretionary income, coupled with the variability of incomes, means that the poor are extremely reluctant to take on new debt or diversify (eg. into cash crops or other enterprises) because of uncertainty regarding their capacity to repay loans or withstand losses.

2.29   Table 2.10 shows the typical consumption pattern of a poor family in the rural terai, although it does not differ markedly from that of other poor households (see Annex II.2).

Table 2.10: Monthly Consumption Pattern of Poor Households
(1984/85 Rs. per month)

	Household (Rs.)	Per Capita (Rs.)	Per Capita /a (US$ Equivalent)
Grains and Pulses	414	58	$3.25
Spices, Vegetables, etc.	87	12	$0.68
All Other Foodstuffs	53	7	$0.42
Sub-total: Food	554	77	$4.35
Fuel, Water, etc.	52	7	$0.41
Clothing	56	8	$0.44
Housing Costs /b	18	2.5	$0.14
Education and Health	24	3	$0.19
Transport	7	1	$0.06
All Other	38	5	$0.30
Sub-total: Non-Food	195	27	$1.54
Total	749	105	$5.89

a/ Converted at Rs. 17.8/US$ (1984/85 exchange rate).
b/ Excludes imputed rental value of own home.

2.30   Of note are the overwhelming preponderance of food and fuel (81% of the total) and the almost insignificant level of consumption of any other goods and services, except clothing. The lack of almost any discretionary purchasing power is reflected in the very low absolute levels of expenditure for other items - even those often considered essential (eg. US 19 cents per capita per month on health and education, six cents on transport).

Food Security

Conditions of the Poor - Summary Table - Food Security				
	RURAL NEPAL		URBAN NEPAL	
	Terai	Hills	Terai	Hills
Calorie Consumption /a	-5%(+25%)	-7%(+15%)	-5%(+7%)	-11%(+3%)
Share of Cereals in Total	86%(80%)	81%(76%)	86%(78%)	82%(69%)
Share of Calories Self-Produced	62%(81%)	71%(80%)	43%(46%)	42%(31%)

NOTE: Corresponding figures for the non-poor shown in parentheses.

a/ Relative to estimated minimum daily requirement (terai 2,140 kcal, hills 2,340 kcal).

2.31   The poor consume, on average, just below the estimated minimum caloric requirements, although there are quite wide variations among the poor and across geographic areas. More disaggregated data show the bottom third of the poor consuming substantially (20-25%) below the required minimum. The problem is accentuated by poor retention of available calories due to parasites and other infections (see Chapter 5 for a fuller discussion of food security issues).

Table 2.11: Food Sources and Consumption Among the Poor-Rural Terai
(Kcal/person/day)

	Market Purchased	Home Produced*	Bartered	Total
Grains, Cereals and Pulses	200	1,233	406	1,838 (90%)
Oils and Fats	24	13	1	38 (2%)
Fruits and Vegetables	35	45	0.5	80 (4%)
Meat, Fish and Eggs	4	5	0.1	10 (0.5%)
All Other Foods	31	44	2	77 (4%)
Total	294 (14%)	1,340 (66%)	409 (20%)	2,043

* Includes food received free.

2.32   Table 2.11 shows the composition of food consumption for a poor terai family; again it is not significantly different from that of other poor households. Such a family subsists almost entirely on rice and dal - a lentil-based gruel - they consume less than 5% of their calories in the form of vegetables and fruit, and almost never eat meat, fish or eggs. During a substantial portion of the year (generally the three or four month pre-harvest period) they will eat only one meal a day.

2.33   Rural families are strongly motivated to grow their own subsistence needs, particularly basic cereal crops. It is likely that this tradition arose in the hills and mountains where isolation and poor access produce very high transport costs which, in turn, make self-sufficiency in food grains more economically rational than in an economic environment in which goods can move more freely. However, the poor rural families discussed in this section do not have sufficient land to produce their own calorie needs, although growing food crops has first claim upon family land and labour resources. Note that such poor families sell only about Rs. 60 (about US$3.50) worth of agricultural produce per month on average (Table 2.5). Non-poor families, with more land resources, can not only produce a larger share of their own subsistence needs, but can also supply their needs by purchases, given that they generate cash income from agriculture of about Rs. 300 - Rs. 400 per month. In summary, poor rural families are not only more dependent upon non-farm sources for food, but also for earning the income (in cash or kind) to supplement what they can grow themselves. Even so, their food intake tends to fall short of their food needs, not to speak of their appetites.

```
       Conditions of the Poor - Summary Table - Living Conditions

                                   RURAL NEPAL            URBAN NEPAL

                                 Terai      Hills       Terai      Hills

Access to Latrine                1% (7%)    6% (19%)   11% (38%)  46% (74%)

Access to Water                 91% (94%)  60% (69%)   93% (97%)  77% (88%)

Persons per Sleeping Room        3.7 (3.0)  3.7 (2.8)   3.8 (2.8)  3.5 (2.4)


Note: Corresponding figures for non-poor shown in parentheses.
```

2.34   <u>Housing Conditions</u>. The vast majority of households, including very poor ones, own a homestead plot together with a dwelling. There is thus no evidence of mass homelessness in Nepal, even among the poor - although housing conditions tend, of course, to be primitive throughout. For all poor families, the average number of persons per sleeping room is just below four.

2.35   The main problem associated with housing conditions is lack of access to water and sanitation. Lack of access to any nearby source of water is mainly a problem in the rural hills, with 40 percent of poor families having to fetch water from long distances. In addition, the drinking water source for all households is usually contaminated. There is almost complete lack of sanitation in rural areas, and very incomplete coverage in urban areas. In the rural terai and hills 99 and 94% respectively lack effective access to a latrine. Almost 90% of poor urban families in the terai have no access to a latrine (54% in the hills). The absence of toilet facilities for the poor produces generally low standards of sanitation, which affect non-poor and poor alike.

2.36   Time Use. The patterns of time use for men and women as well as boys and girls, are presented in Annex II.2 and summarized below. The main impression which these data give is that the people of Nepal live lives of largely unrelenting work.10/   They are not poor because they work an insufficient number of hours but, rather, because the rate of remuneration per unit of time worked is very low.

```
        Conditions of the Poor - Summary Table - Time Use
              (Average work burden - hours per day)

                        RURAL NEPAL          URBAN NEPAL

                    Terai       Hills      Terai       Hills

Adults - Male       8.0 (7.6)   7.9 (8.0)  8.4 (7.6)   7.6 (6.9)
       - Female     9.6 (9.1)  10.5 (11.0) 9.5 (9.0)   9.9 (8.6)

Children Age 10-14
       - Male       3.8 (2.8)   3.8 (4.4)  2.7 (2.2)   2.2 (1.5)
       - Female     6.5 (5.4)   6.8 (7.7)  5.7 (4.4)   5.6 (3.6)


Note:  Corresponding figures for the non-poor shown in parentheses.
```

2.37   Overall, both poor men and poor women tend to work longer hours than their non-poor counterparts. Their work burden is consistently higher, except in the rural hills, where the poor tend to work for slightly shorter hours than the non-poor. One reason for this might be the lack of work opportunities on the miniscule farms of the poor in the hills, combined with the lack of access to nearby off-farm employment opportunities.

10/ The apparent contradiction with high underemployment rates (para. 2.21) is explained by the fact that there is substantial seasonal variation in workloads, by the under-reporting of subsistence activities, and by the fact that in the absence of productive alternatives, many subsistence tasks are probably carried out in a time-consuming fashion.

2.38   Women spend much more time than men on subsistence activities and domestic work. As a consequence the work burden of adult women exceeds that of adult men by about 25 percent. This gender discrimination is common to all household categories, poor and non-poor alike, rural and urban, hills and terai. Gender patterns are established early in life. Girls 10 to 14 years old have a work burden about double that of boys in the same age group, and the pattern seems not to be dependent upon the poverty or otherwise of the family. Partly in consequence, girls spend less than 60 percent of the time that boys spend in education and reading. As noted below, however, girls participation in education _is_ related to poverty, because in all four geographic categories, girls from poor families spend less time in education than girls from non-poor ones.

2.39   <u>Education</u>. The poor lag behind the non-poor in all categories of school enrollment, with the disparity increasing at each higher level. In the rural terai, for example, there is less than one chance in three that a poor child will be enrolled in primary school; for a child from a non-poor household the chance is rather better than one in two. At the same time, children of school age are disportionately concentrated in poor families. Systematically, educational chances are worse for girls than for boys, in the terai than in the hills, in rural areas than in urban and, from poor families than from non-poor ones. The relationships between education and poverty are explored more fully in Chapter 5.

Conditions of the Poor - Summary Table - Access to Education				
	RURAL NEPAL		**URBAN NEPAL**	
	Terai	Hills	Terai	Hills
School Enrollment				
Primary	30% (53%)	53% (67%)	38% (63%)	64% (69%)
Secondary	13% (30%)	12% (30%)	22% (42%)	25% (49%)
Ratio of Girls to Boys in Primary School	0.3 (0.6)	0.5 (0.7)	0.5 (0.7)	0.7 (0.9)

Note: Corresponding figures for non-poor shown in parentheses.

2.40   <u>Health</u>. Among poor families there is continuing malnutrition and illness, and it is almost a certainty that at least one, and often two, children will die as a consequence of ill health aggravated by malnutrition. Health status is universally poor in Nepal - due to the poor coverage of health services, contaminated water supplies, overpopulation, and poor hygiene - it is not therefore primarily income-related. However, the poor exhibit a number of characteristics which make them, on average, less likely to be healthy - including poorer nutrition, larger family sizes, and less access to water supplies, sanitation, and health services. Issues of health care and the poor are discussed in Chapter 5.

III. THE DETERMINANTS OF POVERTY

A. The Dynamics of Poverty - Output and Growth Issues

3.1   The combination of very low average income levels and a fairly uniform distribution of income suggests that the basic cause of poverty in Nepal is excessive population concentrated on an insufficient economic base, rather than the inegalitarian distribution of available wealth. The weakness of the base stems from natural disadvantages which hinder modern sector growth (eg., a difficult topography and location); the early stage of development, manifested in an unskilled workforce and weak managerial capacity; and a failure, thus far, to transform agricultural productivity in the way that the green revolution has done in most developing countries. All of these factors have been exacerbated by sustained rapid population growth.

3.2   Nepal is by nature a high-cost economy. It is land-locked, and suffers from difficult (and expensive) internal communications as a result of extremely mountainous terrain. Among the world's poorest countries are most of those which are land-locked. They suffer from higher border prices for imports, and are at a competitive disadvantage with respect to exports, due to both higher transport costs, and the higher costs of imported intermediate goods. There are of course counter-examples (eg., Switzerland, Austria), but these tend to be located at the center of efficient transport networks, surrounded by diversified high-income economies, and have also compensated by specializing in high value-added and skill-intensive goods and services. Nepal, at the moment, has none of these compensating advantages.

3.3   Approximately a third of the population live in areas in the hills and mountains which are inaccessible by road, where the costs of physical inputs are prohibitive, and marketed outputs are uncompetitive; and where it is infeasible to deliver developmental or social services at reasonable cost. The costs of road-building in these areas is many times that of constructing them in the neighboring plains of India. Unlike other developing countries, Nepal does not have significant reserves of cultivable virgin land to be settled. Net cultivated land represents only 18% of land area, with very limited scope for expansion. Finally, the long open border with India severely limits the scope for pursuing independent strategies to maximize growth (see para. 3.13). These constraints all limit the prospects for alleviating poverty through rapid growth.

Population and the Resource Base

3.4   The most fundamental factor contributing to poverty in Nepal has been the rapid increase in population - which approximately doubled in the last 20 years, and will double again by 2020. Population growth of 2.7% p.a. has eroded the limited gains which have been made in GDP and food production. If population growth had been contained to 1.5% p.a. over this period (instead of 2.7%), then real per capita GDP would have risen 45%, rather than the 14% which was achieved.

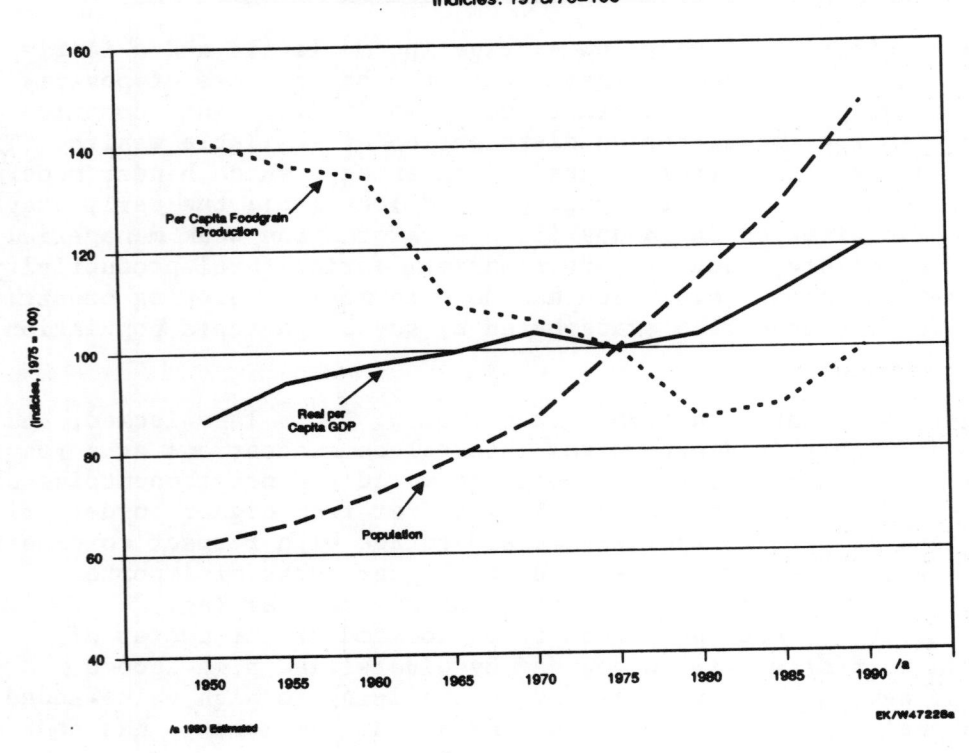

Figure: 2 Effect of Population Growth on Income and Food Availability
Indicies: 1975/76=100

/a 1990 Estimated

The further doubling of population is an almost inevitable result of the momentum of past growth. Even if effective birth control were instituted today, it would take about 15 years until the population started to stabilize, since there are 50% more girls below the age of 14 <u>already</u> <u>born</u> than there are total women currently of reproductive age. There is little effective family planning (only 17% of women practice contraception), and the total fertility rate, at about 6 births per woman, is the highest in Asia.<u>1/</u> The two most critical consequences of this population explosion are unsupportable labour force growth, and saturation of the agricultural land base.

3.5   Population density is now about 6.2 persons per hectare of cultivated land - ranging up to 10 persons per ha. in parts of the hills. Similar levels are found in the fertile Aseatic deltas - whereas in Nepal these densities obtain mostly on drylands of mediocre quality. The density per unit agricultural land has increased by nearly two-and-a-half times over the last two generations - resulting in rapid land subdivision and fragmentation. For example, the availability of agricultural land has declined from 0.6 ha. per person in 1954 to 0.24 per person in 1990. The average farm size in the hills in now below 1 ha. (and less than 0.5 ha.

<u>1/</u>   Population programs, and issues related to the poor, are discussed in Chapter 5.

for 50% of holdings) which is insufficient to support the average farm household of 6 persons.

3.6   As a consequence of this pressure cultivation has been expanded up hill slopes and on to poorer land on plateaus and ridges. Marginal productivity is thus falling, and the land base itself is deteriorating - as soils are depleted, and erosion accelerates with encroachment onto more fragile slopelands. Furthermore, forest cover has been depleted - reducing the availability of fodder, natural fertilizer, and fuel - further decreasing the support capacity, and increasing erosion. A recent study estimates the probable costs of resource degradation at between $320-560 million per annum by 2010, in the absence of remedial measures.[2]

3.7   The response to saturation of rural economies elsewhere has been rapid urbanization. In Nepal the terai has provided a temporary safety valve. Following the eradication of malaria and subsequent land colonization, an estimated 1.2 million persons migrated from the hills to the terai in the 1960's and 70's.[3] However the terai too is now approaching saturation, and further growth will need to be absorbed in urban areas and in non-agricultural employment (see para. 3.20).

3.8   It must be recognized that the terai is a strip of land only 30 miles wide at its widest. There are at most a further 400,000 ha.[4] of forest land which might potentially be converted for agriculture. Even if all of this were converted to farmland (a proposition of dubious ecological soundness), then at existing rates of natural increase and immigration, the rural terai would be saturated within 20 years (with densities of 8 persons per ha. - equivalent to China).[5] HMG's current policy is to restrict further settlement and protect remaining forests. This does not appear to be strictly necessary, although the implications of such a strategy are not well understood, and the Government needs to undertake an analysis of the economic and environmental costs and benefits of conversion of terai forests to agriculture (see para. 3.22).

---

[2] Environmental Resources Ltd. (ERL); *Natural Resource Management for Sustainable Development*, 1989. Costs are estimated in deficits of food, fuel and fodder.

[3] 1961-1981 based on differentials between natural and observed increases.

[4] Equivalent to 20% of currently cultivated area.

[5] See ERL: even with full intensification of agricultural productivity, the carrying capacity would not expand beyond another 5-10 years' worth of population growth.

Economic Performance

3.9   Overall GDP growth has averaged about 3.4% p.a. over the last 20 years[6]/ - these gains have been largely eroded by population increases, with real per capita GDP growing at only 0.75% p.a.[7]/ The distinguishing characteristics have been sluggish growth in the agricultural sector, and limited but steady improvement in the formal sector (average real growth of 4% over the last 20 years), fueled by an acceleration in manufacturing output in the late 1980's. This high rate is, however, on a very low base, and manufacturing still accounts for only 6% of GDP and 2% of employment.

3.10   The changing structure of the economy reflects the early stages of transition out of subsistence agriculture, coupled with rapid growth of the Government sector (Table 3.1). The share of agriculture in GDP has declined from 65% to 52% - but this change is accounted for mostly by the growing share of indirect taxation rather than growth in services and non-manufacturing industries - which have maintained fairly constant sectoral shares (25% and 9% respectively).

Table 3.1 GDP Composition and Growth

	Sectoral Shares				Average Rate of Growth	
	1965	1975/76	1985/86	1987/88	1965-1985	1980-87 /a
Agriculture	65% /b	65%	53%	52%	1.6%	3.9%
Manufacturing	3%	4%	5%	6%	4.4%	5.7%
Other Industry	8%	6%	9%	9%	3.0%	5.5%
Services	23%	23%	27%	25%	3.1%	4.7%
Indirect Taxes	1%	2%	7%	7%	2.7%	4.4%
Total /c	13.5	17.3	24.6	27.7	2.7%	4.4%
US$ per capita	$146	$150	$151	$172		
Population (millions)	10.4	13.0	16.9	17.8	2.4%	2.7%

a/ Real growth rates outside of agricultural based on a composite GDP deflator, which may disguise inter-sectoral variations. Revised data for the last two years suggest a large increase in 1987/88 (about 8%) followed by a decline in 1988/89 (as a result of the Trade and Transit dispute). 1988/89 per capita GDP is estimated at US$164 equivalent.
b/ Proportions based on current price data.
c/ Constant price data including indirect taxes, 1974/75 Rs. billions. 1965 data may not be strictly comparable with later years.

6/ National income data should be treated with caution; the data base is weak and estimates are not consistent across years.

7/ 1970-1989.

3.11   Performance in the agricultural sector - which almost exclusively determines the incomes of the poor, has been even worse. Between 1975 and 1988 foodgrain production increased by about 20%, while population increased by about 40%. Nepal has gone from being a food exporter to a net importer, and cash crops remain of minor importance - accounting for only about 15% of crop production. Yields of almost all key crops have been declining, with the limited increases in output accounted for mostly by expansion of cropped area into more marginal lands. The reasons for this failure have been well documented elsewhere[8], and include:

- inadequate availability of key inputs (fertilizer use, for example, is the lowest in Asia);

- dependence on rainfed agriculture and the lack of effective irrigation;

- the lack of developed marketing networks;

- the lack of location-specific technologies - especially for the hills; and,

- generally ineffective support and extension systems.

The failure to expand agricultural output has a triple effect on the poor, since:

- they are dependent for the majority of their income on their own production of foodgrains.

- it reduces overall food availability, to which they are most vulnerable[9]; and,

- it reduces potential labour absorption on the farms of others.

The relationships between agriculture and the poor are explored in Chapter 4.

Macroeconomic Conditions

3.12   Poverty in Nepal is not, for the most part, a consequence of macroeconomic imbalances. The key features of the economy are illustrated in table 3.2. The balance of payments situation is structurally weak, with a heavy reliance on foreign assistance to finance essential imports. For a small economy (and thus one more likely to be trade dependent) Nepal has low trade/GDP ratios: exports are only 6% of GDP and imports 20%.[10]   Despite

---

[8] See for example: Nepal Agricultural Sector Review; The World Bank, 1990.

[9] Because they are consuming only the minimum requirements to start with; and because it drives up food prices.

[10] 1987/1988.

some recent expansion of manufactured exports (mostly carpets and garments), the value of total merchandise exports amounts to only $10 per capita; the 1989 WDR shows only four countries with a lower level.

Table 3.2: Nepal - Structural Economic Indicators

	1982/83	1986/87	1987/88*	1988/89*
GDP Growth /a	-3.0%	2.7%	9.7%	1.5%
Exports/GDP	3.4%	5.2%	6.1%	5.6%
Imports/GDP	18.8%	18.9%	20.5%	21.8%
Current Account (US$ m)	-$217 m.	-$202 m.	-$267 m.	-$296 m.
Government Revenue/GDP	8.4%	10.3%	10.8%	10.4%
Government Expenditure/GDP	20.7%	19.9%	20.8%	24.3%
Budget Deficit/GDP	-12.3%	-9.6%	-10.0%	-13.9%
Debt Service Ratio	-3.9%	-6.9%	-8.8%	n/a
Inflation	14.2%	13.3%	11.0%	8.9%
Exchange Rate /a	-4.1%	-7.9%	-2.7%	-13.6%
Interest Rate /b	14%	17%	17%	17%

* Revised Preliminary Estimates.

a/ Change from previous year.
b/ Average of commercial lending rates.

Sources: HMG; Economic Survey of Nepal; and World Bank Country Economic Memorandum, various tables.

3.13   The principal macroeconomic distortions arise from Nepal's situation vis-a-vis India. Being locked behind India's highly protectionist trade barriers, and integrated into a market dominated by administered prices, Nepal is dealing with second-best solutions in which distorted Indian prices are the effective border prices for most goods. The main effect of this is that the rates of return on deflecting third country goods to India are sufficiently high, and risk-free, that there is little incentive for entrepreneurs to invest in real-sector productive enterprises. Furthermore, Nepal's capacity to follow independent macroeconomic policies is extremely limited. Any attempt to liberalize without adopting the Indian structure of protection would result in large-scale smuggling from third countries. Even supposing such a scenario were politically feasible, the resulting appreciation against the Indian rupee would hurt the tradeable sectors much more than any gains from liberalization. The main effect of these distortions on the poor is the dampening impact on overall economic growth, and especially on employment-creating growth.

3.14   Within this framework of second-best solutions, the main economy-wide prices, the exchange rate and the interest rate, are not badly out of alignment. Nor does Nepal face an insupportable debt-service burden, having generally followed prudent borrowing practices, and enjoying sustained inflows of concessional resources.

Structural Adjustment

3.15   Unlike many other countries which have initiated structural adjustment programs, Nepal at the time it did so was not greatly out of macro-economic adjustment, given the inflow of concessional resources it experienced and anticipated. The main outstanding distortions were removed under a stabilization program initiated in 1985 with support from an IMF stand-by arrangement. The structural adjustment program and two structural adjustment Credits in 1987 and 1989 were therefore concerned primarily with accelerating growth through better economy-wide and sector management.

3.16   As was analyzed in Chapter 2, the rural poor (who constitute about 95% of the poor in Nepal) participate only partially in the monetary economy. They are, therefore, largely insulated in the short-term from changes of the kind envisaged in the structural adjustment program. In the longer term, to the extent that the program is successful in accelerating growth and, in particular, in improving the delivery effectiveness of government programs, the poor will benefit from it. There are some components, however, which may have a direct impact upon the poor. These include the following:

    (a) Improvements in the procurement and distribution of fertilizer. These changes are unlikely to adversely affect small farmers; however most of the poor are not in a position to effectively use fertilizer, and the reforms have not been sufficiently far-reaching for poor farmers, particularly in the hills, to gain significantly.

    (b) Measures to improve the financial viability and targetting of subsidies by the National Food Corporation (see Chapter 6), which, if implemented, could potentially help the poor.

    (c) The Program involves a more participatory approach to forestry and irrigation management and proposals for the privatization of smaller irrigation schemes. These are changes in directions likely to benefit small farmers.

3.17   The first two adjustment credits have, correctly, concentrated mainly upon improving general economic and development management. As such, they have had, or are likely to have, limited direct impact on the mass of the poor. Certainly adjustment in Nepal has not included the kind of cuts in public services or increases in the prices of staples which are often associated with adverse impacts on the poor under structural adjustment in other countries.

The Trade and Transit Dispute

3.18   The short-term economic environment changed dramatically in March 1989 when the trade and transit (T&T) treaties with India expired. The most immediate effect was the lack of inputs and shortage of fuel which shut down many industries and disrupted transport. It seems likely that the effect on the poor of the T&T dispute has not been particularly pronounced, since they exist largely outside the cash economy and the urban areas where the shocks were felt most strongly. The longer term impact has been to slow the momentum of economic growth, driving up prices and costs, delaying construction of productive infrastructure, and further eroding confidence of potential investors in the formal sector. While these constitute a medium-term set-back, the impact on poverty in Nepal over a horizon of twenty years or more is unlikely to be significant. The T&T dispute is the most recent illustration of the inherent vulnerability of the Nepalese economy as a result of its position vis-a-vis India - from the perspective of poverty alleviation, it is this which is the significant factor, rather than the dispute itself.

Emerging Trends

3.19   The fundamental problem is the pressure of continuing, rapid population growth illustrated in Figure 3. The population is expected to grow to about 24 million by 2000, and to 32 million by 2010. In the absence of a very aggressive population program it will expand to about 47 million by 2030.

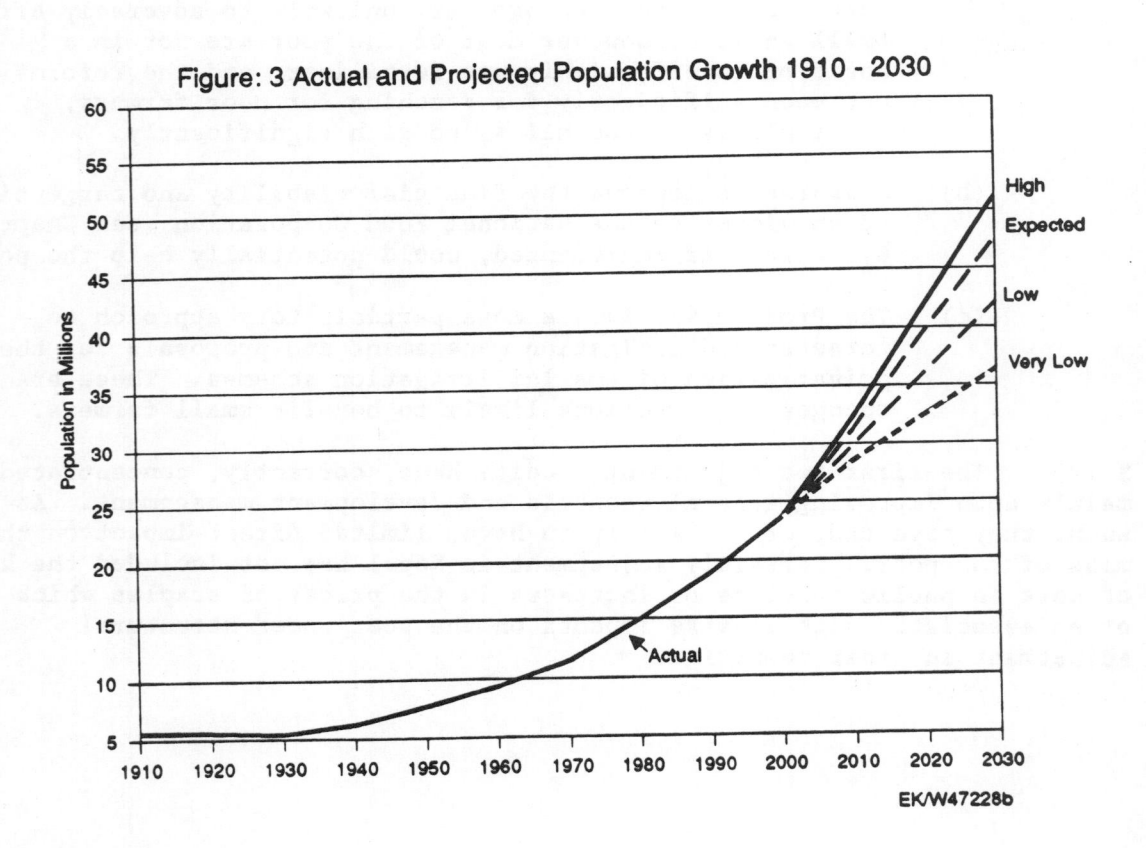

Figure: 3 Actual and Projected Population Growth 1910 - 2030

The demographic profile is particularly worrying - there are for example, 40% more children in the 0-4 year-old cohort than there are in the 10-14 year-old cohort. This has disturbing implications for future incomes, employment trends, and saturation of the land base - especially in the terai.

3.20   Figure 4 shows the projected spatial distribution of population over the next 20 years.

Figure: 4 The Changing Spatial Distribution of Population
(Population in Millions)

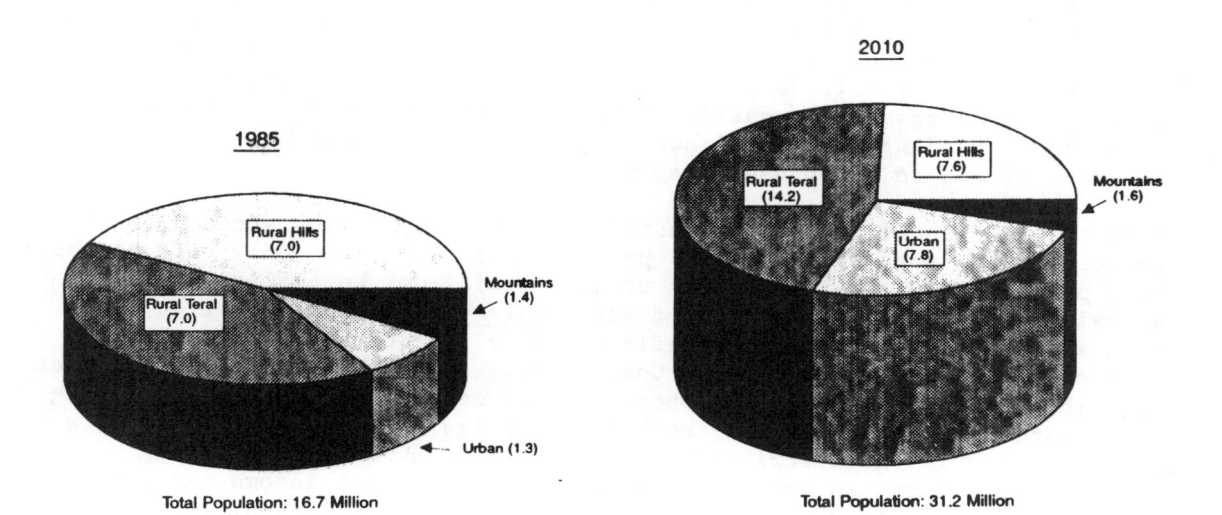

1985
- Rural Hills (7.0)
- Rural Terai (7.0)
- Mountains (1.4)
- Urban (1.3)

Total Population: 16.7 Million

2010
- Rural Hills (7.6)
- Rural Terai (14.2)
- Mountains (1.6)
- Urban (7.8)

Total Population: 31.2 Million

Even with a conservative limit on population density, the rural population of the terai will double from 7 million to 14 million by 2010; and the urban population will increase more than five-fold - to almost 8 million people. This means, among other things, that cultivated land availability in the terai declines from an average farm size of 1.35 ha. per household to just under 0.75 ha. per household - implying the need for a doubling in agricultural productivity if current household incomes are to be maintained.

3.21   With respect to the massive increase in urbanization the concern is that with only limited industrialization, the agricultural base may not expand rapidly enough to create sufficient secondary off-farm employment opportunities for this new landless population. The likelihood is that there will emerge in Nepal a large underclass of urban poor who will not have the social fabric (however tenuous) which currently provides some support to the rural poor.

3.22   In the 20 years or so that it will take to effectively curb population growth, the task confronting HMG will largely be one of managing this transition - from the hills to the terai, and into urban areas. In the case of the terai, it will require a detailed analysis of the scope for conversion of remaining forest lands, along with a program to manage their conversion; and taking steps to expand agricultural productivity to increase the carrying capacity of the terai, including tenancy reform and irrigation improvements (see Chapter 4). It is recommended that HMG undertake analytical work on how best to handle the movement of incremental population growth to the terai and to urban areas - including the investment and institutional requirements, and the costs and benefits of opening up further terai lands.

3.23   The labor force will double over the next twenty years (from 6.7 million in 1985 to 13.6 million by 2010). 11/ In the past the economy has absorbed marginal population growth though a combination of smaller farms, the opening of the terai, and labor market expansion. With the saturation of arable land almost all of these new entrants will have to be absorbed in off-farm employment activities - a prospect of staggering proportions. By the year 2005 they will be joining the work force at the rate of 8,000 per week, posing a tremendous challenge to the Government and the economy. The scope for expanding sectoral labor absorption and the measures needed to do so are explored in subsequent chapters.

3.24   Even under the best expected conditions (i.e. sustained high GDP growth and low population growth) per capita income could only rise to about $270 p.a. by 2010 12/ This sort of growth will not be sufficient to alleviate poverty on a significant scale unless such growth is tightly focussed in areas which will benefit the poor. (a strategy for such "balanced" growth is explored in Chapter 7). Furthermore, the current flat distribution of income may not obtain in the longer run, due to increasing urbanization and monetization, both of which are usually accompanied by a deterioration in the distribution of income, and because with rising labour/land ratios, an increasing share of agricultural income can be expected to go to landowners, and a decreasing share to labourers.

3.25   Projections show that under expected conditions of growth the poorest 40% would remain below the poverty line by 2010 - even if the current distribution of income can be preserved. If the distribution of income were to deteriorate to a level more typical of developing countries, then even with moderately strong (3.3%) GDP growth the lowest 40% of the population would drop well below levels necessary for survival (e.g. US$40 per capita annually for the lowest quintile, US$65 for the second lowest).13/ Even under a scenario of rapidly rising per capita GDP, and no erosion in the distribution of income, the poorest 20% would remain in absolute poverty for at least the next 20 years, leaving a minimum of 6 million absolute poor by 2010.

11/ Labor force growth over the next 20 years is largely deterministic, since new entrants have mostly already been born. See Social Sector Strategy Review (SSSR) for underlying assumptions.

12/ Constant 1987/88 Rs. equivalent - Source: SSSR. Assumes sustained real GDP growth of 4.3% p.a.

13/ See SSSR for detailed projections and assumptions.

3.26   The policy conclusions that emerge are two-fold:

  (i)   that under all conditions there will be a need to design programs to provide relief to the poor on a continuing basis for at least the next 20 or 30 years; and,

  (ii) for growth to have a substantial impact on reducing the number of poor, policies must be followed which guard against a deterioration in income distribution, by pursuing growth strategies which favor the poorer groups in society.

B. The Mechanics of Poverty - Distributional and Equity Issues

The Social Context - Inequality at the Village Level

3.27   While by international standards most Nepalese are uniformly poor, localized studies point to large perceived disparities between poor and wealthy at the village level. This is so because the distribution of assets is more obviously skewed than that of income, and because at the margins of survival small absolute differences in income (say $50 per capita p.a.)14/ mean the difference between near-constant hunger, and a moderately comfortable surplus.

3.28   Distinct discontinuities exist between groups at the local level. In the hills the distinction is often between those producing surplus food, and those producing less than about six months' supply. (Indeed poverty in Nepal is popularly defined in terms of the number of months' food a family produces). In the terai it is between, at the top, landlords, landowners (who may or may not be poor), tenants, and at the bottom, the landless. With respect to location, those in the valleys are generally the better off, those on slopes and ridges less so, and those in more remote areas are usually the absolute poorest, with the exception of some mountain groups which have traditionally depended on trading.

3.29   Across all ethnic groups and regions there has been a tendency on the part of the poor to see the constellation of civil servants, local political leaders, and landlords as being a homogeneous group, who have at best little in common with the rural poor. These perceptions, the poor reach of public services to rural areas, and the difficulty in communicating between officials and illiterate peasants, have often resulted in a wide gulf between the rural poor, on the one hand, and the local political process and Government services on the other.

3.30   At the village level a range of microeconomic and social factors affect the incomes of the poor, including land tenure, labour market and debt relations, and the political environment. The following sections investigate the role of various constraints, the extent to which they prevent the poor from increasing their incomes, and measures which might be needed to lift these constraints.

14/ The difference between being 20% below the poverty line associated with minimum caloric requirements, and 20% above - and thus firmly in the 'non-poor' group.

Land Tenure and Tenancy

3.31   Nepal does not exhibit the concentration of landholdings typical of many developing countries. This is largely because farm sizes have been reduced to a low average level by population growth and land fragmentation. However, data suggest that the top 5% of owners control about 40% of cultivated land, while the bottom 60% control about 20%. This distribution appears skewed, particularly from the perspective of a marginal farmer, because access to a small plot (say less than 1 ha.) - especially of poor quality - is not generally enough to produce an above-poverty income[15], whereas a "large" landowner in Nepal (say 5 ha. - which would be small elsewhere in the world) produces a substantial surplus. The operational distribution of holdings is summarized in Table 3.3.

Table 3.3: Size Distribution of Operational Holdings

Size of Holding Hectares	Terai Average Size	Terai Percent of Holdings	Terai Percent of Land	Hills Average Size	Hills Percent of Holdings	Hills Percent of Land
<0.5	0.09	45.4%	2.8%	0.18	54.1%	11.6%
0.5-1.0	0.74	13.0%	6.5%	0.75	18.7%	16.3%
1.0-3.0	1.77	27.7%	33.1%	1.49	22.1%	38.3%
>3.0	6.15	13.8%	57.7%	5.67	5.1%	33.8%
Total	1.17	100.0%	100.0%	0.86	100.0 %	100.0%

Source: IDS, The Land Tenure Systems of Nepal, 1986. As with all such data, the proportion of land in the largest holdings is probably under-reported.

3.32   Tenancy and share-cropping arrangements are widespread, although it is particularly difficult to obtain accurate information concerning the extent of tenancies, partly because many households both rent-in and rent-out land, and partly because many tenancies are unreported because landlords have reacted defensively to the threats to their interests contained in the 1964 Land Reform Act (see discussion below). While officially reported tenancy dropped after the 1964 land reform, (to 8% of farm households in the 1981 agricultural census) informal tenancy went underground and flourished.

---

[15] This is because on unirrigated land cropping choices are limited and the poor cannot take advantage of their high person-land ratio to get high returns per unit of land. See M. Lipton, Land, Assets and Rural Poverty. World Bank, 1985.

A sample survey found that somewhat over 21% of farm families in the hills, and 39% in the terai were tenants; for an average of 31%.16/

3.33   Almost all tenancies are arranged on a share-cropping basis. Rent payable to the landlord is usually half of gross produce.17/ In general there is no sharing of inputs, at least on registered tenancies. Informal tenancy is apparently more participatory than certified tenancy, with landlords making some contributions to farm and entrepreneurial inputs.

3.34   Assessing the degree of landlessness is difficult because of inconsistent reporting and definitions. The census reports less than 1% of households being landless, but other sources report figures of between 10 and 20%. It seems clear that the 45 percent of holdings in the smallest size class in the terai (Table 3.3) with an average size of 0.09 ha., include very many effectively landless households. These have a homestead block rather than a "farm". As an approximation which is consistent with the data, it would be safe to assume that almost all rural hill families have some agricultural land (although by no means enough to produce their subsistence requirements), and about 20 percent of rural terai households are effectively landless.

Land Reform Efforts

3.35   Under the 1964 reform ceilings were placed on landholdings (ranging from 3 ha. in the Kathmandu valley to 18.4 ha. in the terai), "excess" land was to be purchased from landlords and redistributed - mostly to sitting tenants or neighboring farmers. In the end about only 1.5% of cultivated land was distributed among 10,000 peasant families.

3.36   The reform was accompanied by a tenant registration drive (although only an estimated 10-20% of tenants were certified). An enormous problem was that the law provided for "voluntary" surrenders by tenants; this provision was exploited to ease out tenants. In addition, land offices demanded payment of enormous extra-legal fees, and record-keeping did not allow easy identification of large landowners' holdings. Those tenants who did get registered, and who learned their rights and stuck to them (a very small minority) probably benefited from the reform. However, to the extent that the reform achieved the objective of greater tenancy security, the agricultural system was rendered less flexible, and the land market less viable.

3.37   Even though little land was in fact redistributed, there remains a strong sense of uncertainty. Under the current system those with "registered" tenancies can potentially claim title to the land they till.

16/ IDS, The Land Tenure System in Nepal.

17/ Ibid, p. 53.

Because of the fear of possible transfers to tenants (either in completion of the 1964 reform, or under some new reform initiative) landlords, tenants, and would-be landlords are all using land inefficiently and inflexibly.

3.38   The current tenancy system is inequitable, and perhaps more significantly, it is inefficient. One commentator has remarked that it would be difficult to design a system _less_ conducive to maximizing production. As a result of uncertainty and the perceived threat of transfer to tenants:

    (i)   landlords are unwilling to make productive fixed investments (eg. in irrigation);

    (ii)   tenants are rotated frequently (removing the incentive for _them_ to make improvements);

    (iii)   output is shared, but not inputs - so that tenants equate private returns with one-half of marginal output, and there are thus incentives to produce below the optimal level;

    (iv)   small plots are farmed less intensively than is optimal using insufficient family labour, rather than risk using tenants;

    (v)   landlords prefer to use Indian labourers who will not make claims on the land; and,

    (vi)   tenants and smallholders lack land titles (due to the difficulties of registration) which they need for loan collateral or to participate in irrigation programs.

Conclusions

3.39   In answer to the question: Is the current tenancy system inequitable? The answer is probably yes. Can much be done about it? Probably not. Since the problem is one of excess labour and a shortage of land the net terms of payment to landholders will, one way or another, be determined by the scarcity of land - whatever regulatory framework is put in place. Furthermore, the incentives to evade tenancy reforms are great, and where the peasantry is not organized, and the administrative capacity of the government is weak in its reach to the community level, the ability to abolish tenure types or regulate factor shares through policy is ineffectual at best.

3.40   There is some scope for physical redistribution in the terai,[18]/ and the distribution of operated land can potentially have more direct

---

[18]/ A redistribution of the 20% of terai land in the largest holdings, if broken into small parcels (0.5 ha.) and effectively targetted, could possibly raise some 400,00 households out of poverty.

equity effects than tenancy reform, but: (i) redistribution would provide only a temporary respite, and (ii) there has to date been little sign of political support for a redistributive reform. An arithmetic example illustrates the first point: if all land is excess of 5.4 ha. 19/ were redistributed, this would bring 47% of those in the lowest landholding-size class up to 1 ha., and would not benefit the landless at all. Furthermore, there are few reliable income-producing assets other than land. The absence of alternative investment instruments and social security measures act as a powerful disincentive both for landlords to sell land, or for middle-class support for a land reform.

3.41   With respect to efficiency issues, it is evident that there are many problems with the present agrarian structure. First, there is unfinished business left from the 1964 Land Reform Act, with registered tenants operating land at low efficiency levels. Second, the perceived threat of agrarian reform and the possibility of giving security to tenants has led to defensive action on the part of landowners. Third, there is already a large and growing class of landless, many of whom might become tenants if the laws were changed. Fourth, there are innovations, such as shallow tube-wells, which hold the potential for substantial production increases, but which are not reaching small farmers on account of tenure and collateral difficulties.20/

3.42   For the reasons discussed above, a redistributive land reform does not seem to currently be a prospect in Nepal. However, some steps can be taken to regularize land tenure and tenancy arrangements, to reduce uncertainty and thus to remove some of the barriers to efficient use of land, especially in the terai. The key is de-linking tenancy reform from land redistribution. Elements of such a reform might include: (i) bringing the 1964 reform to closure; (ii) giving legal status to alternative tenure arrangements which do not necessary involve full security of tenure; and (iii) streamlining procedures for land registration and transactions.

19/ The ceiling proposed in the 1986 IDS Land Tenure Report.

20/ There are also some special problems of land management in the vicinity of rapidly growing urban areas, such as in the Kathmandu valley. However, it is recommended that these land policy issues on urban fringes be treated as a separate problem from policy towards rural land proper. In this way, issues of agrarian reform can be considered in their own right and separate from issues of urbanization policy.

3.43   While it is clear that these type of changes are needed, substantial thought needs to go into designing a workable reform given the complexity of the issues, and the incentives for non-compliance. The tendency in Nepal is to frame reforms in terms of abolishing tenancy and transferring ownership to tenants. We do not know the extent of tenancies, the reasons for which they exist, (including the extent to which they may give the system economic flexibility it would not otherwise have), and the extent to which the country's rural social safety net is to some extent embedded within tenancy. We therefore do not know whether the existing tenancy system in fact serves some useful purpose for which other policies would have to be devised if tenancy were to disappear. <u>It is recommended</u> that analytical work be undertaken to (a) collect data on the extent and type of non-formal tenancies; (b) assess their effect on equity and efficiency; and (c) thereafter design a tenancy and agrarian reform package which at a minimum removes the barriers to efficient use of land contained in the current system.

<u>Labour Force Issues</u>

3.44   The central features of the labour supply and demand landscape in Nepal are rapid labour force (LF) growth on one hand, and on the other the absence of the sort of agricultural transformation (eg. through irrigation) which would substantially affect labour absorption in farming; coupled with relatively slow growth of employment in other sectors.[21] Between 1971 and 1981 the labour force grew by 41%, whereas the total population grew by only 28%. This is the result of population growth being concentrated in the lower cohorts from which new LF entrants are drawn, and of a 5% p.a. growth in female participants - fueled by changes in the terai - where female LF participation rates doubled from 16% to 31% in ten years.

3.45   The labour market is overwhelmingly dominated by agriculture and by self-employed subsistence activities. According to the 1981 census, about 86% of the work force were self-employed, of whom 97% were engaged in agricultural pursuits. To the extent that they are employed outside of subsistence agriculture, we have seen that the poor are engaged overwhelmingly as paid agricultural labourers (30% of the economically active poor in the terai), or as casual labourers - mostly in construction (about 10% in the hills).

3.46   While most households participate in the labour market, the proportion of time hired out is low (19% for men, and only 6% for women, according to one survey). Estimates of open unemployment are low (about 6%), however time-use estimates of <u>underemployment</u> are generally in the range of 45%-65%. There is naturally substantial seasonal underemployment in rainfed agriculture (even adjusting for the high subsistence work burden), and except in some areas, at some times of the year, there is very

---

[21] The expected response would be to bid down wage rates, although it is difficult to see the potential for this in Nepal where wages appear to be at (or close to) a subsistence level.

little in the way of off-farm work. MPHBS data estimates about 40% of available family labour days go unutilized, and one survey found, for example, that casual agricultural labourers are able to find employment only 68% of the time that they seek it.

Employment Arrangements

3.47   In many parts of the hills, where the lack of a cash economy is combined with low mobility of labour (due to difficult access), one-to-one labour exchange between families has traditionally functioned in place of a wage labour market. In the terai, on the other hand, labourers are employed for payment in grain, usually under long-term contract, but also on a daily wage basis. There are two types of contracts: attached labourers work exclusively for one landlord, while semi-attached labourers must work for the contracting landlord when he needs them, but are free to work elsewhere at other times.

3.48   Both are usually contracted for a period of one year at an agreed price. Labourers are paid a subsistence wage or slightly below in grain and foodstuffs, but they also get access to credit, sometimes use of livestock, and often use of a very small piece of land on a sharecropping basis. These arrangements assure the farmer an adequate supply of labour during the peak season, and a fixed income (however meagre) to the employee. Such employees are often provided with a loan sufficient to carry them through the agricultural year. In the western terai there exists a form of bondage whereby such loans may bind the poor in a form of serfdom - in some cases indistinguishable from slavery - to a given landlord.

3.49   Non-agricultural employment of the poor is generally in construction or, to a much lesser extent, cottage industries. There is little documentation of the labour arrangements under which such labour is engaged. However anecdotal evidence suggests that there are a number of exploitative practices, especially in construction.

3.50   These reportedly include not paying wages, advancing pay in the form of meals at inflated prices, and charging excessive contractors margins at the expense of payments to labourers. While these practices reflect in part the reality of low real wage rates; they are exacerbated by existing contracting arrangements and the lack of power of the poor to effectively demand the wages due to them.

Wage Levels

3.51   In 1989 average wage levels were about Rs. 23 per day in the hills, and about Rs. 17 per day in the terai.[22] The following table provides a good indication of current wage rates:

---

[22] These are an average of daily wage rates for agricultural labour and unskilled construction work - the relevant rates for the poor.

Table 3.4: **Estimated Current Daily Wage Rates (1989)**
(Rs. per day)

		Men (Range)	Women (Range)
Agricultural Wage:	Terai:	Rs. 15	Rs. 13
	Hills:	Rs. 23	Rs. 20
District Wage: /a	Terai:	Rs. 20 (18-22)	Rs. 17 (15-20)
	Hills:	Rs. 31 (22-42)	Rs. 26 (16-42)

a/ Supposedly paid on all public works

Source: Mission Interviews (November, 1989)

3.52   **Sufficiency.** There is a remarkable uniformity across surveys of daily wage rates at around 4.25 kg. of paddy equivalent. Adjusting for days worked, and dependency ratios, this comes to about 1850 calories daily per capita - about 15% below the estimated minimum requirement. [23] In terms of cash income, these rates are equivalent to per capita incomes[24] of Rs. 2130 per annum in the hills ($76 p.a. in 1989 US dollars) and Rs. 1575 in the terai ($56 p.a.); well below the NPC defined poverty line.

3.53   This suggests two things: that wages have been bid down fairly uniformly to close to a subsistence minimum; and that at prevailing wage rates and family sizes the poor are not going to rise out of poverty by getting access to employment alone.

3.54   **Wage Trends and Variations.** Several studies document declines in real wages of about 30% over the 1970's, however there are no consistent time series, so it is difficult to accurately assess movements over time. Compiling data from a variety of sources suggests that the average wage was equivalent to about Rs. 22/day in the early to mid-70's, and about Rs. 15/day in the mid-1980's (in constant 1987/88 prices) compared to about Rs. 18-20 today. The best one can conclude from the inconsistent data is that

---

[23] Assuming 185 days worked per year, a dependency ratio of 2.0, 50% conversion to food, and 3,450 cal./kg. of rice.

[24] Making the same adjustment for days worked and dependency ratios.

real wages have fallen during the last 15 years, recovered somewhat, and are currently either stagnant or declining.

3.55   There are marked regional and local variations in wage rates. Nominal wages in the hills seem to be consistently about a third higher than in the terai (consistent with a subsistence minimum, adjusted for grain price differentials). There are in addition very wide local variations in unskilled wage rates (e.g., ranging from 20 to 50 rupees per day in the hills) which result largely from remoteness (with higher cost structures and lower mobility of labour), and closeness to urban areas, where wages may be higher due to the level of agricultural development, proximity to markets, and the availability of alternative employment opportunities. The nominal daily rate for female workers is 15-20% below that for men, although some of this differential is accounted for by differences in hours and type of work.

****

BOX: TYPICAL INCOME OF A WAGE EARNING FAMILY
- Terai Agricultural Labourers -

The family is contracted to a small landowner, the father works 7:00 a.m. until 6:00 p.m. about 15 days per month for which he is paid 3 kgs. of paddy per day, in addition to a breakfast and lunch equivalent to 1.5 kgs of rice. The mother might work about 10 days per month from 11:00 a.m. till 5 or 6:00 p.m. for 3 kgs. of rice. In addition, they have the use of a small garden plot (about 0.08 ha.) belonging to their employer, which may yield about 180 kgs. of rice annually in income. There is likely to be an adolescent child or older relative who works about the same amount as the wife. In addition, the family may get about 7 days a month of off-farm work at Rs. 15 per day. One child may tend livestock - either their own or the employer's. In the off-season, or on unemployed days, they may catch mice or crabs in the fields of landholders, or fish in local streams.

Total Family Income:	Rs. 683 per month
Family Size:	7
Per Capita Income:	Rs. 98 per month (US$3.45 equivalent)
Food Consumption:	1700 kcal/person/day Rice and other cereals and pulses account for 90%

****

Labour Supply/Demand Issues

3.56   The high estimates of underemployment, the reported declines in wage rates, and village-level anthropological studies, all suggest that labour is available in surplus. In the hills out-migration has largely balanced labour force growth; but in the terai a class of landless agricultural labourers has emerged which exceeds the demand for labour on average - although not necessarily during periods of peak demand.

3.57   The labour force is growing at about 3.3% p.a. on average (5.5% in the terai). Some of this labour is being absorbed in agriculture; however a number of contradictory forces influence the rate of absorption:

>   Firstly, there is a farm-size effect. The reduction in farm-size due to subdivision increases labour intensity per unit of land, but the smaller farms tend to be worked exclusively by family labour - so the hiring in of landless (and marginal) labourers - who tend to make up the poor - is reduced.
>
>   Secondly, there is a cropping intensity effect. The shift to irrigation and high yielding varieties, and the use of fertilizer, all result in more intensive production and use of labour - especially in the terai, and especially among large farmers.
>
>   Thirdly, there is a land tenure effect. With the threat of transfer to tenants, landholders have reverted to self-cultivation, thus requiring more labour supply, while at the same time previous or potential tenants have become wage labourers. At the same time current tenancy arrangements discourage optimal production, and thus reduce the absolute quantum of labour used.

3.58   The increased demand for labour as a result of intensified cropping and the return to self-cultivation has not pressured wages upwards. Instead as a result of population pressure, all (or almost all) of the increased rents due to agricultural improvements have gone to landowners. Barring further major advances in technology (eg. the spread of effective irrigation), with LF growth of over 5% p.a. in the terai we should expect real wages to be pushed down further.

3.59   In the non-farm sector, the effect of urbanization is to open up additional employment opportunities, although the rate of growth of the urban labour force (about 20% p.a.) probably more than offsets the income effect for the poor. The monetization of the economy also has contradictory labour demand effects. It creates opportunities and increases the volume of off-farm activities (although often not those which the poor engage in); and at the same time it has reduced demand for the skills of traditional occupational castes, whose products are substituted for by cheaper imported goods. It appears that the availability of off-farm employment opportunities has, so far, had little impact on labour force factors, (or on wage rates) - because in aggregate there exist relatively few such opportunities. This is however changing slowly in areas with greatly improved access, where studies suggest that the proximity to markets and alternative sources of income may have had a positive impact on agricultural wage rates.

3.60   A number of things can be done to improve labour absorption. Perhaps the most important step would be to get irrigation working in the terai. This, with subsequent cropping intensification, would result in demand for about another 500,000 person-years of labour annually, about a

50% increase.25/ Opening up remaining terai forest land - if feasible - could contribute about another 500,000 full-time equivalent jobs,26/ and regularizing the land tenure situation would also have an impact. Improving access in the hills would enhance labour mobility, as well as lifting constraints on non-formal sector growth and cash-cropping which could contribute to substantial absorption.

Employment and the Poor

3.61   Because they do not have sufficient land, the poor are disportionately dependent on labour market incomes. The level of such incomes depends on four factors, which may be subject to varying degrees of policy intervention: (i) family composition; (ii) participation rates; (iii) employment levels (the number of hours (or days) worked per participant); and, (iv) wage levels. The poor in Nepal are adversely affected by each of these factors, both singly and in combination with one another.

3.62   Analysis of <u>family composition</u> (Chapter 2) shows that the poor have larger average family sizes and higher dependency ratios than the non-poor, so that those who work in poor families have to support more non-working family members.

3.63   There is downward pressure on <u>participation rates</u> among the poor, not just because of family composition but because of a number of compounding factors. <u>Firstly</u> due to their lack of land, the costs of participation are higher (including in particular the costs of temporary migration). <u>Secondly</u>, illness and disability limit their capacity to participate actively. <u>Thirdly</u>, they have to spend more time in gathering fuel and fodder, and to a lesser extent water. <u>Finally</u>, females, of whom there are a greater proportion in poor families, exhibit lower participation rates due to cultural prohibitions, as well as frequent pregnancy and childbirth.

3.64   <u>Un/Underemployment.</u> Available data shows unemployment tends to be higher among the poor (by as much as 50%) in rural areas. The fact that the poor have less (or no) land to work with implies that in the absence of a robust demand for labour they are also more likely to be underemployed. Micro-level studies in the terai indicate that it is the landless labourers who work the least number of days per year. Finally, caste and ethnic group membership still have a major influence on access to employment - both of which militate against employment of the poor.

3.65   <u>Earning rates</u> for the poor are also lower than average, both for on-farm and off-farm activities (see Chapter 2). This is due to: (i) the lack of assets (eg. land) with which to work; (ii) lower levels of education and skills; (iii) the relative lack of political leverage to influence wage

---

25/ Assuming full double-cropping of rice and wheat, and a shift of about 35% of currently rainfed land to irrigation, see Appendix II.6.

26/ Assuming full irrigation and intensive cropping - see Appendix II.6.

rates; (iv) the fact that they tend to live in more remote areas where agricultural productivity is lower and wage rates are less influenced by the presence of off-farm opportunities; and (v) the higher proportion of children and female labour force participants, who are employed in lower wage rate occupations.

3.66   The fundamental facts of the labour market are excessive population on a limited economic base - which is keeping labour incomes at the margin close to survival levels. Within this framework the scope for increasing labour incomes is necessarily limited. However, there are a number of areas in which improvements are possible.

3.67   <u>Labour Absorption</u> First, there is a need to expand employment creation in off-farm activities. An increased program of labour-intensive public works is one promising area (see Chapter 4); however given the scale of the problem the Government's contribution will be limited mainly to enabling mechanisms, by improving physical access and infrastructure, removing petty restrictions, providing credit and other incentives for employment-generating activities. Careful design is needed to ensure that these measures open up opportunities for the poor rather than the non-poor (eg. by focussing on the non-formal sector, construction, etc.).

3.68   Secondly, <u>measures to improve agricultural productivity</u> can potentially have a major impact on labour absorption, probably increasing it by about 50% in aggregate (para. 3.60). Equally importantly, agricultural employment growth of these dimensions would have a major impact on the demand for off-farm services and labour.

3.69   The scope for policy interventions to affect <u>wage rates</u> is limited - because they are bound to be driven by marginal agricultural productivity, and heavily influenced by the free flow of labour from India. However, the following would have some impact at the margin:

- enforcement of District Wages for public works would serve not just to transfer more resources to the poor, but also - if the volume of works is large enough and their regularity assured - to bid up the reservation wage in selected areas;

- improved physical access tends to increase the range of off-farm options, and thus bid up local wages - although this may affect the poor less than others;

- programs involving group formation, credit to the poor, and other empowerment measures have been shown elsewhere to improve the bargaining power of the poor vis-a-vis employers, allowing some bidding up of wages.

All of these measures will be likely have more effect in the hills than in the terai, where wages will tend more to be determined by the Indian level.

3.70   With respect to agricultural employment contracts, in answer to the question: Are they exploitative? the answer is probably yes. Can much be done about it? probably not, given the shortage of land and surplus of labour; although measures which empower the poor (eg. education, group formation, and decentralization, if effective) can serve to improve their influence on the terms of contracts. Construction contracts may be another matter. Monitoring and enforcement has shown to have some impact on the payment of wages and reduction of exploitative arrangements. Similarly the use of smaller local contractors and community-organized works has tended to lead to more wage income for the local poor. Both of these areas deserve further investigation to determine what does and does not work (see paras. 4.42-4.44, and 6.61-6.64).

Migration

3.71 There has always been temporary migration in Nepal, however it has taken on a new importance and complexity in the last 25 years, with the population explosion, and opening up of the terai. Migration is basically of three types:

(1) Seasonal (3-6 months), during the agricultural slack season in the hills;

(2) Medium-term (up to 3 years), mostly to India, to supplement incomes and savings; and,

(3) Permanent - as the result of land loss, indebtedness, and lack of employment opportunities.

3.72 Seasonal migrants generally go to the terai, and medium-term migrants to India. Internal migration is almost exclusively from the rural hills to the terai and urban areas, while international migration is more complex, involving a network of flows both into and out of India.

3.73 Recently short-term migration has taken on much more of a distress nature - as remaining plots of land cannot provide adequate food to support families, and thus the males outmigrate in the slack season so that both they and the family members left behind can adequately feed themselves. Anecdotal evidence suggest that remittances from such migrants are very small or non-existent.

3.74 Long-term temporary migrants (over 6 months) go overwhelmingly to India, where they work in the cities as watchmen, or in Assam and the border areas as construction labourers, porters at coal depots, or in forestry. They tend to stay about two years, and do bring back some cash savings. A substantial proportion (perhaps 30-50%) of hill males undertake some form of temporary migration - although the relative importance of long vs. short-term migration is not known.

3.75 There exists no good aggregate data on the importance of remittances in incomes of the poor. In financial terms it seems that migration is more of a coping strategy than one for poverty-alleviation. The Rastra Bank estimated that only 8% of the cash income of rural hills households came from remittances; and one survey suggests that the relevance of remittances is lower for the poor than the non-poor. Temporary migration is none the less an important coping strategy - especially in some areas, and for some families. One sample of three Districts found 25-30% of households to be "solely dependent on seasonal migration for supplementary income" (i.e. presumably meaning that they had no other source of off-farm income).

3.76   Permanent Migration. It is difficult to accurately quantify the scale of internal permanent migration, but estimates suggest it accounts for about half of the marginal population growth in the hills. It is not clear how the income effects of permanent migration are distributed. Studies suggest (not surprisingly) that it is the relatively poor who out-migrate, and that on arrival in the terai most end up as marginal landless labourers. Thus they go from a completely untenable situation (in the hills), to a not much better one in the terai - an improvement, but certainly not a strategy which will raise them out of poverty.[27]

3.77   Data on the scale of international migration is inconclusive, although there is no doubt that the labour market in the terai (and elsewhere) is affected by substantial inflows of Indian labour. The only quantitative analysis (based on census data) suggests that net international migration is approximately zero, but there is probably significant under-reporting of both in- and out-migrants.

3.78   The pattern of international migration is complex. The 1964 land reform inadvertently encourages landholders to use Indian labourers and tenants rather than hill people. At the same time wages in some Indian border areas may be lower than in the terai, at least seasonally. This has resulted in a paradox - that hill people go to India, while Indians migrate into the terai.

3.79   Nepal has attempted to reduce the inflow, but to little effect - both because restrictions are physically difficult to enforce, and because there exist strong economic incentives for non-compliance. It is difficult to see what Nepal can do in this regard. The use of obvious restrictive measures which are successful in other countries would invite counter-vailing measures from India. The regularization of tenancy arrangements may go some ways towards removing incentives, but would be unlikely to have a dramatic impact.

Debt and Indebtedness

3.80   Almost everyone in Nepal is in debt. The average size of borrowing appears to be in the range of Rs. 1,150 p.a. [28] representing about 6% of annual income on average - but the absolute level of borrowing is fairly uniform across income classes - so that the poor are borrowing up to 20% of their incomes annually. The volume of borrowing from institutional sources has been growing rapidly, but this growth has been concentrated almost exclusively among the non-poor; less than 10% of institutional credit is estimated to go to the poor.

---

[27] The income effects of migration are not well documented, it is recommended that further research by undertaken on them.

[28] 1984/85 from MPHBS expressed in 1988/89 prices.

3.81   Institutional credit still accounts for only about 20% of total borrowing (and almost none among the poor). Private moneylenders appear to account for another 20-25%, while the majority of borrowing is from other non-formal sources. These range from friends and neighbors to large informal credit societies, and cover a spectrum of arrangements from small interest-free loans, through debt-bondage to landlords or employers, to large-scale commercial financing (up to Rs. 50,000) by highly structured local credit groups, which function in lieu of formal financial institutions (see tables in Annex II.7).

3.82   The majority of borrowing by the poor is not for investment purposes, but rather to sustain consumption. Surveys show that between 60% and 90% of borrowing by small farmers is for consumption (including marriages, catastrophic health care expenses, etc.), and that less than a third is for capital investments. It is worth noting that small farmers are not using credit on a seasonal basis to finance agricultural operations in the way that would normally be expected (only 10-20% of borrowing among small farmers). This is probably because (a) they are not in a position to use cash inputs, and (b) they are averse to risking additional debt on the basis of outcomes which depend on (highly uncertain) rainfall.

3.83   Interest rates vary from 10% to 150% per annum. On most informal borrowings they lie in the range of 30-40%. In addition there is often a transaction charge, which is reportedly higher for poor and marginal farmers (10-25% of the principal) than for large farmers (5-10%). Institutional interest rates are in the range of 15-20% (as low as 7.5% on subsidized agricultural credit for selected investments). However numerous sources report that the actual costs of using institutional credit may be substantially higher, due to high transaction costs in terms of time and extra payments (often 10% of the amount borrowed), which are usually higher for the poor than the non-poor. One study estimates that the effective rate is close to the 35% p.a. charged by non-formal sources.

3.84   Village studies often report that chronic indebtedness contributes significantly to impoverishment of the poor. However, because debt relations are often combined with employment and tenancy arrangements, and repayment is often in kind (in either labour or grain), it is difficult to quantify the extent to which debt service at high interest rates erodes the incomes of the poor. Calculations suggest, however, that debt service on typical amounts borrowed might amount to a minimum of Rs. 400 p.a. [29] - or 15% of discretionary income among the poor. More importantly, the MPHBS data show that those in the lowest deciles are accumulating debt at a staggering rate, and repaying at only about a fifth the rate that they are borrowing. This is clearly an unsustainable situation. While it almost certainly reflects some under-reporting of repayments, it is reinforced by reported net decreases in assets among the lower deciles. Other surveys also confirm that the poor are consistently borrowing to cover their consumption shortfall.

---

[29] In 1988/89 Rs. per household, based on MPHBS data.

3.85   The exploitiveness of debt relations varies greatly across communities - influenced as much by degree of social cohesion as by economic factors. They range from:

(i) systems which involve temporary debtor/creditor relationships between social and economic equals -- a kind of "revolving" indebtedness in a relatively closed and socially homogeneous community.

(ii) systems which express and maintain economic (and sometimes social/ritual) inequality and may involve expropriation of labor, grain and sometimes the use of land -- as well as the expectation of political support from the debtor -- but which are essentially equilibrium systems. In other words, the dependence of the debtor upon the creditor is also a source of security for the debtor and the relationship does not involve permanent transfer of assets to the creditor.

(iii) systems which result in a permanent net transfer of assets (e.g. land) sometimes leading to bonded labor or forced migration out of the community for the debtor.

There is no conclusive evidence on the importance of the various types of indebtedness, but local studies suggest that there is both loss of land (and especially usufructuary rights) as a consequence of debt, as well as continuing instances of bonded labour, in which the debtors' body (or that of his wife or offspring) is effectively seized in payment of debt.

3.86   In answer to the question: Are current debt relations exploitative?, the answer is: It depends on the type of debt. Can anything be done about it? Among the poor, only partially. The main problem is insufficient incomes to meet consumption needs. In this context massive debt is a symptom - not a cause - of poverty, and the solution to the debt trap lies not in providing more or cheaper credit but in raising incomes. However, some welfare improvements are possible by increasing access to credit on better terms than moneylenders (provided the credit is not captured by the non-poor), and some programs have had success in doing this (see Chapter 6). In addition providing alternative credit sources to the poor may in some instances break the cycle of dependence on landlords and employers - allowing them to negotiate better employment or tenancy conditions.

Gender and Poverty

3.87   Women are at the center of poverty alleviation strategies in Nepal - because they undertake a very large share of household agricultural production; because they are disproportionately not participating in education, and their lack of participation in education is the main contributor to low overall educational status among the poor; and finally because they are central to changes in fertility and hygiene behavior. Furthermore, in Nepal and elsewhere experience has shown that some of the greatest family welfare impacts among the poor can be achieved by placing more income in the hands of women.

3.88   Women predominate slightly in poor households, although it is not clear if this is cause or effect (i.e., because a family is poor for some other reason (like lack of land) it has more females in it (due eg. to out-migration of the males), or the family is poor because it has a higher proportion of females in it, eg. due to lower participation rates). The proportion of female-headed households, on the other hand, is no higher among the poor than the non-poor (para. 2.24).

3.89   Women's earnings are substantially lower than men's - mostly due to lower participation rates, because of the household division of labour, which assigns domestic work mostly to women. Cultural and social customs also limit female participation in work outside the home - especially in the terai and especially among certain ethnic groups/castes. Wage levels for the same work - adjusted for hours and type of work - appear to be slightly lower for women - however, there are substantial pressures with respect to health, education, mobility, and job-entry which keep women in lower wage work.

3.90   Women as a group are generally poor because they are in poor households, rather than because of gender-specific discontinuities in the distribution of income. However within poor households, which are subsisting on average at just about survival levels of food consumption, they are consuming somewhat less than equal shares, and they are thus more vulnerable to declines in incomes. Although anthropological studies report that that boys receive preferential care and feeding when resources are scarce, a recent survey of intra-household food allocations showed a fairly even distribution between male and female children. In all families, however, daughters-in-law reportedly receive a lower food share, and cultural practices prescribe less food for women in pregnancy.

3.91   The security of women's income is less than that of men - both because they lack individual ownership of assets (especially land), and because their LF participation is more tenuous. Furthermore, there exist a number of ways in which the welfare of women is probably substantially below that of males in families at the same income level, viz.:

- women work significantly more hours per day, especially fetching water, fodder, etc;

- their caloric requirements may be higher due to the longer work day, pregnancy, and lactation;

- they suffer more illness as a result of frequent childbirths;

- they are less likely to receive medical care or education; and,

- in the hills, they are the ones who are left behind to survive off a non-producing farm in the off-season (although there is no data on the consumption effects of this).

There is scope for substantial productivity increases by:

(a) increasing the labour force participation of females - mostly through education and fewer childbirths;

(b) freeing up time spent on domestic tasks - particularly fetching water, fuel and fodder, by providing water supplies, reforestation, and trail improvement;

(c) ensuring that agricultural services address the needs of women farmers (eg. by training of extension workers, who otherwise seldom deal with female farmers; the Lumle research station has had some success in doing this); and,

(d) by pursuing programs which support on-farm income-earning activities which put more income in the hands of women (for instance, the Production Credit for Rural Women program).

Social and Political Constraints

3.92   Nepal is in a state of transition, mirroring the evolution of the economy from subsistence agriculture to a more monetized one. The traditional political system, which prevailed until the middle of this century, was highly paternalistic. It involved a complex web of feudal patron-client relations; the majority (and certainly the poor) were not expected to participate actively in it, nor particularly to benefit from it.

3.93   Traditional patron-client relations are breaking down as a result of increased monetization, off-farm employment opportunities, and education. However, in many cases they are being replaced with revised versions, which substitute cash payments for kind; multiple ties with one patron (eg. a landlord) with single ties to several people (eg. civil servants, businessmen, moneylenders). The modern political system has generally mirrored the traditional power structure; representing largely the same interests, although there is a complex intermingling of traditional political forces with economic interests, and cultural and caste groupings.

3.94   There is a widespread perception that at least until recently the political system was designed to operate for the benefit of a very small minority. The system had employed a range of formal and non-formal pressures to discourage unmanaged change or participation by the rural peasantry. The moves toward decentralization and more participatory approaches over the last five years represent changes in the right direction. The more dramatic recent changes (in early 1990) appear to reflect an acceleration of this trend, although it is too early to tell how deep rooted the changes are, and whether the political evolution will be accompanied by any corresponding economic restructuring.

3.95   As in most countries a whole array of social and political pressures make it difficult for the poor to effectively take advantage of services or developments which could benefit them. These include an aversion among the poor to approaching official institutions, the tendency of officials to prefer to work with the better off, and an unwillingness or inability to deal with illiterate peasants. In addition, the private costs to the poor of using such services (in terms of time, and extra-legal payments) can be very high.

3.96   The process of decentralization thus far does not appear to have been seriously geared to addressing this problem. The poor *can* more effectively demand and utilize services when they form into groups, and some programs have had some success in doing this. However there has been a tendency to confuse such groups with emerging political forces, (which had been banned), or as a threat to the established distribution of power. In fact some NGO and donor groups report incidents of physical intimidation when such group formation activities are undertaken. The implication is that measures designed to empower the poor (eg. in project design) will only succeed to the extent that the larger body politic is prepared to accept such change. While the environment may be more open than it has been, there may remain serious constraints which limit the likely effectiveness of such measures.

IV. POVERTY AND THE PRODUCTIVE SECTORS

A. Agricultural Incomes and Poverty

Impact of Agricultural Growth on the Poor to Date

4.1   The three major crops in Nepal are rice, wheat and maize. Rice is the preferred staple and is grown throughout the terai and in some areas in the hills. Maize is the dominant crop in the hills. Wheat production, in the hills and terai, has increased rapidly both in terms of area and yield. Millet, barley and potatoes are grown on poorer soils and at higher elevations. Most farmers grow small amounts of pulses and oilseeds (mustard, linseed and rape) for their own consumption. Sugarcane, cotton and jute are grown as cash crops in the terai - although the poor grow almost no cash crops.

4.2   Production of staples has increased only 1.4% per year over the past 20 years despite the expansion of cropped area at the rate of 1.9% per year. Yields of most crops have been stagnant or have declined with expansion of cropping onto marginal land and decreasing fertility due to increased cropping intensity and low fertilizer use. This slow growth has occurred despite an increase since 1975 of 980% in area irrigated, 140% in area covered by high yielding varieties, and 240% in annual sales of fertilizer.

4.3   Local improvements in agricultural productivity have occurred for rice, wheat and maize, but only in areas with better soils and/or irrigation and access to chemical inputs. Case studies show that the poor tend to be located in places with less fertile soils,[1] this limits their capacity to increase output, or to adopt productivity improvements which may be feasible for other farmers. (In many cases, of course, their poverty is at least in part a consequence of their poor quality landholdings). In addition, the poor have less access to irrigation and less ability to pay for chemical inputs. While the poor do grow rice, wheat and maize, because of geographic location and soil quality they are more likely than wealthier farmers to grow millet, barley and potatoes, all crops that have not even experienced localized increases in productivity.

Effectiveness of Existing Programs in Reaching the Poor

4.4   It must be stressed at the outset that general agricultural programs are intended more to promote overall growth than to benefit the poor specifically. The following sections, however, focus in particular on their impact on the poor, in keeping with the objectives of this report. An assessment of their effectiveness more generally has been provided in the World Bank's recent Agricultural Sector Review. Government agricultural programs have faced many difficulties and have had limited impact, but to the extent they have benefited any farmers, the poor have received fewer benefits. This is in part because many of the poor have so little land that

---

[1] See, for example, Pachico (1980), and McDougal (1968).

they cannot be effectively reached by general programs. It also reflects some inherent bias in the structure of general programs, which have not adequately taken account of the constraints faced by poor farmers, as well as a natural tendency (largely correct) to focus efforts on larger farmers where the greatest productivity gains can be made. Moreover, agricultural programs have with some justification concentrated on the terai, where potential productivity is higher and constraints fewer. However, the poor are disproportionately located in the hills and mountains. The following sections briefly describe the impact of the general agriculture programs on the poor; programs with a more specifically poverty focus are discussed in Chapter 6.

4.5   Irrigation. Government irrigation programs have historically concentrated on building large-scale, gravity-feed systems in the terai. The poor have benefited when they have owned land in the command area and when water delivery has been efficient. There is little evidence that the poor are systematically excluded - except to the extent that they are less likely to have title to land. Large-scale irrigation projects are difficult to target at the poor except by giving priority to poor areas. This has not been done explicitly, although large irrigation systems have been built in poor sections of the western terai. Most importantly, however, the large systems have severe water delivery problems, limiting benefits to poor and non-poor alike.

4.6   The focus of the government irrigation program has recently shifted from large-scale schemes to smaller-scale farmer managed systems. The government is in the process of handing over control of small and medium-scale irrigation systems to farmers in order to improve performance. In addition, it is trying to improve the operation and maintenance of existing public systems and, once water delivery is improved, improve the collection of water user fees. Under the new irrigation sector program, groups of farmers can receive assistance in constructing irrigation systems, or rehabilitating and/or expanding existing systems, provided they meet a set of technical criteria. The current irrigation effort could be tilted towards the poor by giving priority to groups consisting of poor farmers or by concentrating on poorer areas of the country. There is evidence of very high economic returns to investments in shallow tubewells in the terai, but the poor do not have sufficient land, in most cases, to justify the investment based on their own production potential. Work is being done to organize groups of farmers for purchasing and operating small tubewells which would make irrigation more accessible to poor farmers.

4.7   Input Supply. The supply of fertilizers has been controlled by the Agricultural Inputs Corporation (AIC), a government corporation with a monopoly on importing and wholesaling fertilizer. AIC is also involved in production and wholesaling of seeds. The AIC has been criticized for inefficiency, and failure to ensure timely delivery of inputs. Actual demand for chemical fertilizer from poor farmers is probably quite low, given their production systems and lack of irrigation, but supply is inadequate nevertheless. Demand for improved seed is strong, but the AIC is only able to produce a fraction of the improved seed being used. Recent changes in legislation and the establishment of the National Seed Board provide the opportunity to establish a system of private seed production. A pilot project on farmer-operated seed production and distribution has been successful in the hills and could be expanded, allowing AIC to withdraw from seed production altogether. More timely supply and wider availability of inputs would probably be possible through a system of small private traders more responsive to the demands of local farmers.

4.8   Fertilizer use by poor farmers will remain low even if fertilizer is available. In the absence of irrigation fertilizer use is a high-risk strategy. For poor farmers, the risk is compounded by the large cash costs involved (perhaps half of their available cash incomes). Poor households can only finance these costs by taking on further debt which they are ill-equipped to do, especially on the basis of highly uncertain returns under rainfed conditions. Therefore, in the absence of irrigation, poor farmers will continue to use minimal amounts of purchased inputs.

4.9   Research and Extension. The poor, in general, have less access to irrigation and chemical inputs than the non-poor, and lower quality soils. They are also more likely than the non-poor to grow rough cereals such as millet and barley. Poor farmers want crops with stable yields over a variety of conditions, and crops that require few chemical inputs. Until recently, however the focus of the research system has been on varieties of rice, wheat and maize that require irrigation and high levels of purchased inputs. This means that for most of the poor the work done by the research system has been largely irrelevant, either because they do not have irrigation, cannot afford inputs, or grow different crops. There have been some exceptions: the Lumle and Pakribas research centers have undertaken farmer-centered research that seeks to minimize chemical input use. In addition, recent work sponsored by USAID has started to focus on green manuring and bio-fertility. These efforts should be expanded since they are directly relevant to the needs of poor farmers.

4.10   The extension service has been unable to reach poor farmers because its message is irrelevant for many of them, it is not structured to reach them directly, and the socio-economic gulf between poor farmer and extension agent often inhibits contact. The model farmer system which was followed until recently by the extension service used wealthier and more educated farmers to demonstrate new techniques. This system allowed a wider dissemination of knowledge using fewer agents, but it meant that the poor were missed. Evidence from local level studies suggests that the farming systems of wealthier farmers are significantly different from those of poorer farmers, that poorer farmers know this, and therefore give less weight to the demonstration pilots. Extension agents rarely visit poor farmers 2/ , and when there is direct contact the difference in status often prohibits real communication; this is especially true with women farmers.

4.11   The research and extension system has recently been restructured to use a farming systems approach, so that information moves in both directions. The extension system has adopted a "contact group" method to increase the number of farmers directly communicating with extension services. In addition, the research system has recently shifted to a

---

2/ See, for example, APROSC-Baseline Study for Crop Intensification Program (1985).

outreach approach which stresses on-farm crop testing and increases the contact between farmers and researchers. This increased contact has shifted the focus of research towards low input systems and should eventually increase the amount of research done on livestock, which are an integral part of the farming system, and on horticulture, which has the potential to raise incomes substantially in accessible areas. These changes will help the poor provided they are included in, and participate in, the contact groups and provided the research outreach program uses poor farmers to test crops.

4.12   Credit. Increased credit is crucial for providing small farmers access to shallow tubewell irrigation and for providing production loans once irrigation is in place - although it is only the medium-poor who are likely to be in a position to utilize agricultural credit. General institutional credit programs by the commercial banks and ADB/N are not used by poor farmers because, among other reasons, collateral requirements are too high and application procedures too complex. Targeted credit programs for the poor such as the Small Farmers' Development Program (see Chapter 6) have had some success in lending to the medium poor, but transaction costs are high and even where they are in place many poor farmers still do not participate in them.

Potential and Constraints in Hill Agriculture

4.13   What then is the best that can be expected from agriculture in raising in the incomes of the poor? For the following sections we have calculated the maximum incomes the agricultural poor could possibly generate - given their typical landholdings, household sizes, and likely cropping patterns. The analysis is based on the potential yields of each of the major crops under varying conditions (eg. whether irrigated or not, with or without modern inputs, on valley floors or slopes, etc.). It draws on farm budgets and agronomic data in the recent Irrigation Master Plan, and from the World Bank's Agricultural Sector Review (See Annex II.6 for details).

4.14   Recognizing that there is enormous diversity among farm households in the hills, they may still be divided into four groups with different agricultural potential and constraints (Table 4.1). Large and medium farmers (27% of hill farm households) derive most of their income from agriculture, but also rely on other sources in order to produce above the poverty-line incomes. Relatively small improvements in agricultural production would allow these farmers to produce above poverty-line incomes from agriculture alone. Almost three-quarters of hill farm households, comprising about 7 million people, have holdings of less than 1.0 ha. Of these households, about one-quarter are small farmers (about 1.7 million people) with holdings that average 0.75 ha. in size; the remaining three-quarters are marginal farmers (more than 5 million people) with holdings that average 0.18 ha. - approximately equal in extent to two or three house lots in many western countries. Even a doubling of agricultural productivity would not raise them above the poverty line.

Table 4.1: Characteristics of Farm Household Groups in the Hills

	Large	Medium	Small (Poor)/a	Marginal (Very Poor)
Average family size	8	6	5	4
Holding size (ha.)	>3.0	1.0-3.0	0.5-1.0	<0.5
Percent of households	5%	22%	19%	54%
Ave. holding size (ha)	5.7	1.7	0.75	0.18
Share of Household income from agriculture	76%	70%	58%	47%
Share of 1985 poverty line income derived from agriculture	76%	61%	46%	34%

a/ Landholding and household sizes for the poor and very poor drawn from MPHBS data - Household landholding distribution from IDS includes some non-poor households in "small" landholding class.

Source: MPHBS (1989) and IDS (1986), *The Land Tenure System in Nepal*.

4.15   Two spatial factors are important in understanding agricultural potential and constraints: topographical location, (i.e., whether on valley floors, slopes or hilltops); and location with respect to access, i.e., to transport infrastructure and markets. Crop yields throughout Nepal are low, but yields from valley floors are significantly higher than for hill slopes under both rainfed and irrigated conditions. The importance of the relationship between topography and productivity for this study lies in the fact that the majority of the poor in the hills do not farm valley floor land, nor do they have frequent access to irrigation. To be poor in the hills means not only having very small holdings, but also having holdings of lower inherent productivity and more subject to stress or shocks, eg. droughts and landslides, than the non-poor.

4.16   This picture of a general relationship is complicated by the complexity of mountain agroecosystems. Elevation, slope, aspect, and season all combine to determine the thermal regime (length of the growing season), potential evapo-transpiration (water balance), and photosynthetic rate (dry weight increase) of crops. As a result, there is a lot of variation in the potential of small plots, since microclimates can be found in the hills which both pose constraints on production technology and also offer opportunities for specialization, particularly in horticulture.

4.17   Accessibility affects cost and availability of chemical inputs, credit, and information. Reliable data on the proportion of the hill population that is accessible are not available, but it is estimated that up to two-thirds live beyond one day's walk of a motorable road. Access difficulties limit the agricultural productivity of farmers by raising input costs and restricting access to markets. Where markets are poor, farmers reduce risk by using farming systems that maximize stability of production and minimize reliance on external inputs.

4.18   With the introduction of irrigation, and using currently available technologies, farmers can earn more than three times what they now earn from rainfed land. However, even with these substantial increases, the small landholdings of households in the small and marginal categories limit their ability to become self-sufficient from agriculture. Marginal farmers have an average of only 0.18 hectares of land. Even with all land irrigated and access to chemical inputs, they could earn only half the poverty line income from field crops if located on valley floors, and only 34% on slopes (Table 4.2). Small farmers with an average of 0.75 hectares of irrigated land can do significantly better, producing well above-poverty incomes on the basis of field crops alone.

4.19   Potential incomes from switching to tree crops are quite high, between Rs. 25,000 and Rs. 50,000 per hectare, if markets are accessible. With these levels of production even the small holdings of marginal farmers are capable of producing between 50% and 95% of a 1988/89 poverty line income. This represents a significant increase over the income possible from an equivalent area in foodgrains or oilseeds. However, these levels of income for tree fruits are built up only slowly, usually over a 12-year period, and require capital investments equivalent to Rs. 10,000 - 14,000 per holding. For these reasons it is unlikely that the poor in the hills, with their limited assets and high levels of debt, will be able to invest in these types of enterprises without a very high level of assistance. Nevertheless, a horticulture program which provided the necessary extension advice coupled with credit could help a proportion of poor households in accessible areas. Furthermore, many of them could benefit from a partial shift to tree crops and vegetables, which could be introduced gradually with only limited reduction in subsistence grain output (which they are unwilling to give up).

4.20   While potential improvements in productivity are large, most poor farmers will be unable to achieve them. Few poor households have access to irrigation, and even with projected expansion of irrigation in the hills, most poor households will still be dependent on rainfed crops. Improvements of up to 20%-30% in the productivity of rainfed crops may be possible through improved management practices, but, while these are significant increases, they are not enough to raise the small and marginal farmers above the poverty line. These low input farming systems will require careful management of soil fertility using natural fertilizers such as mulch and manure. Even if irrigation is available, increases in productivity depend on increases in chemical inputs which are unavailable in remote areas and only sporadically available in accessible areas. Clearly, agricultural production alone will never be able to raise most poor and very poor hill households above the poverty line, although it can make a significant contribution.

Table 4.2: **Potential Monthly Per Capita Income From Field Crops and Percentage of Poverty Line Income For Poor Households, 1988/89** 3/

	Hills		4/ Terai
	Valley Floor	Hill Slopes	
(1) **Small** (0.75 ha.)	(0.75 ha.)	(0.75 ha.)	(1.1 ha.)
Rainfed (Actual)	Rs. 92 (44%)	53 (25%)	Rs. 76 (38%)
Current year-round irrigation (Potential)	Rs. 298 (142%)	151 (72%)	Rs. 228 (115%)
Future year-round irrigation (Potential)	Rs. 341 (162%)	239 (114%)	Rs. 335 (170%)
(2) **Marginal** (0.18 ha.)	(0.18 ha.)	(0.18 ha.)	(0.5 ha.)
Rainfed (Actual)	Rs. 27 (13%)	16 (8%)	Rs. 39 (20%)
Current year-round irrigation (Potential)	Rs. 89 (43%)	45 (21%)	Rs. 117 (59%)
Future year-round irrigation (Potential)	Rs. 102 (49%)	72 (34%)	Rs. 172 (87%)

Source: Based on **Master Plan for Irrigation Development in Nepal** 1989, See Annex II.6 for yield assumptions.

		Small	Marginal
Family Size Assumptions from MPHBS:	Hills	5.0	4.0
	Terai	7.1	6.3

3/ The calculation of potential income from current year-round irrigation assumes that farmers have reliable access to water and chemical inputs. The calculation of potential income from future year-round irrigation assumes that irrigation water control improves, cropping patterns evolve, and that farmers increase their input use accordingly.

4/ The estimates of income possible from **rainfed** agriculture differ from current incomes in Table 4.1 because: (i) Table 4.1 includes income from livestock and horticulture while Table 4.2 considers only income from field crops; (ii) estimates in Table 4.2 are based on **average** cropping intensity per hectare while small farmers often crop more intensively; (iii) some "small" and "marginal" farms in Table 4.1 have irrigation and are therefore more productive than purely rainfed farms.

Potential and Constraints in Terai Agriculture

4.21 The constraints to agricultural production in the terai are fewer than in the hills, but growth in agricultural output has been only marginally better, 1.5% per annum versus 1.4% per annum. As in the hills, agriculture is characterized by many very small holdings; 58% of households in the terai operate 1 hectare or less. Unlike the hills, about 20% of households in the terai are completely landless and agricultural wage labor is an important component of income for poor households.

4.22 According to the MPHBS the typical poor household has about one hectare of land and produces 40% of the poverty line income from agriculture (Table 4.3). This household will plant all of its khet 5/ land with rice during the monsoon season, using several different varieties, one of which may be a modern variety. It will use little or no inorganic fertilizer and no insecticide on the rice crop because it is too risky given the lack of water control. The household will depend primarily on family labour except during periods of peak demand such as transplanting and harvesting. During the monsoon season the household will plant non-khet land to pulses. During the dry season they will plant half the khet land with maize and leave the rest fallow. For maize they will use no purchased inputs and no hired labour because the maize yields are too low and too variable, depending entirely on residual soil moisture and very sporadic dry season rains. Animal manure is used for fertilizer, but is not sufficient for the amount of land, and in areas where forests are completely gone, animal manure is used for fuel instead of fertilizer.

Table 4.3: Household Income and Landholdings in the Terai by Level of Poverty

	Very Poor	Poor	Non-poor
Family Size	6.3	7.1	6.8
Land Operated (ha)	0.5	1.1	2.8
Household Income from Agriculture (%)	37%	50%	73%
Share of 1988/89 Poverty Line Income derived from Agriculture (%)	15%	40%	127%

Source: MPHBS (1988/89 rupees).

5/ Khet: land suitable for growing rice.

4.23   The very poor farmer has an operational holding of half a hectare or less, and generates only 15% of the poverty line income from agricultural production. The farming system of the very poor farmer is similar to that of the poor farmer with several important exceptions. The very poor farmer will probably not own a team of plow animals because they are too expensive to buy and feed. This will raise the cash demands of production considerably and may reduce the amount of land the very poor farmer is willing to plant in the dry season when the risks of crop failure are higher. The very poor farmers will not use even the small amounts of chemical fertilizer used by poor farmers because they have very little cash income.

4.24   With the introduction of irrigation, earnings per hectare could reasonably be expected to triple through improved crop yields and more intensive cultivation (Table 4.4). Households with irrigated land using

Table 4.4: Potential Income Per Hectare from Field Crops in the Terai and Area Needed for Poverty Line Income, 1988/89

	Estimated Maximum Income Per Hectare	Area Needed for Poverty Line Income (family of 7)
Rainfed (Actual)	5,865	2.8 ha
Current year-round irrigation (Potential)	17,670	0.9 ha
Future year-round irrigation (Potential)	25,945	0.6 ha

Source: Master Plan for Irrigation (1989) and MPHBS (1989).
Poverty Line (1988/89) for the Terai= Rs.197 per capita.
See Table 4.2 for assumptions concerning current and future production with irrigation.

currently available technology could earn a poverty line income from 0.9 hectares, and using foreseeable future technology could earn a poverty line income from 0.6 hectares. Recall, however that 45% of rural terai households own less than 0.5 ha., and 20% are completely landless. Households with access to urban markets could earn a poverty line income from less land, if irrigated, by growing fruits and vegetables. Even so a large number of households would still be unable to survive on own production alone.

4.25   In addition to benefiting from increases in their own production, poor households may benefit from increases in the demand for agricultural wage labor. Presently, poor farmers earn only half their income from self-production, and very poor farmers earn only a third of their income from self-production. Both groups earn an additional 30-40% of their income from agricultural wage labor. Labor inputs increase substantially with irrigation in the absence of mechanization, and irrigated farms tend to use more hired labor. With the projected increases in irrigated area, demand for labor should also increase by as much as 50%.

Aggregate Potential for Agriculture to Contribute to Poverty Alleviation

4.26   About a third of the agricultural poor in the hills have enough land to potentially raise themselves above the poverty line through agricultural improvements - but only if their land is irrigable - otherwise they may be able to produce about half of the poverty-level income on their own land. For the balance - who make up about half the population of the rural hills - their holdings are too small to ever produce more than a third to a half of the poverty-level income from field crops, even under the best of circumstances (eg. with year-round irrigation and full use of modern technologies). While irrigation in the hills could be expanded to some unirrigated areas, most of the poor are located on slopes and/or land that cannot be irrigated - for them agriculture can never generate more than a small share (20-30%) of a minimum family income. The possible exceptions to this would be households in accessible areas that shift completely to horticulture. However, few very poor households will be able to make this shift.

4.27   In the terai, about 15% of the poor have large enough holdings (say about 0.6 ha.) to produce above-poverty household incomes, if fully irrigated. Another 25% have sufficient land to achieve perhaps half of poverty-level incomes from their own production of grains. As in the hills, some of these very smallholders could produce sufficient income by switching eventually to fruit and vegetable production. About half of the rural terai poor have insufficient land to raise themselves out of poverty through their own agricultural production.6/

4.28   A shift from rainfed agriculture to irrigated agriculture would increase labor demand by about 180 days per hectare per year, more if households shift from field crops to horticulture. It is estimated that 500,000 hectares of land in the terai not currently irrigated are irrigable. If all irrigable area were irrigated, including area now forested, then yearly demand for agricultural labor could be as high as 554 million days in the terai and 315 million in the hills. Increases in agricultural production in the hills will not lead to significant increases in the demand for agricultural wage labor. Family labor and labor exchange arrangements should provide the increased labor required.

4.29   Recommendations. While technical recommendations in the field of agriculture are beyond the scope of this paper, it is clear that some general directions are worth pursuing. Firstly, the greatest impact on the poor will come from getting irrigation working in the terai, by

---

6/   Possibly more - there is no accurate breakdown of those owing less than 0.5 ha. in the terai by income level; some, with irrigated land, will be non-poor, many however are effectively landless.

rehabilitating and improving operations of the large public schemes and handing over management of medium-sized schemes to farmers, but most importantly by increasing the spread of small scale irrigation, (especially shallow tubewells) to small farmers. This requires: (a) efforts to form them into groups to achieve economies of scale and overcome fragmentation problems; (b) regularization of land tenure and registration, combined with (c) a package of credit and technical support, such as is being provided under the recent irrigation sector program. In the hills slow continuing expansion of small irrigation works will be helpful, although the beneficiaries are often the non-poor.[7] One option may be to use the selection criteria under the irrigation sector program to give priority to poor farmers in receiving assistance, provided they meet the established technical requirements. Reforestation and community forestry projects also need to be pursued for their impact on the availability of fodder and hence of organic fertilizer.

4.30   Secondly, the current input distribution system is unlikely to reach small farmers effectively. Steps have been taken to remove the remaining Government monopolies and restrictions on inputs, but careful follow-through is required to ensure implementation at the local level. Consideration should be given to actively promoting a network of small private traders who would: (a) have financial incentives to seek out and travel to small farmers; (b) be willing to provide the very small quantities of inputs which they require, and (c) have an interest in providing them with technical advice. Experience elsewhere has shown that such traders provide both inputs and a network for marketing agricultural produce; although they will only serve the accessible areas.

4.31   Thirdly, the recent shift to a farming systems research and extension system should be strongly supported. To benefit the poor in particular the new research outreach program should: (a) place the focus of research on low input systems with stable yields; (b) promote more work on livestock, which are an integral part of poor farmers' farming systems; (c) encourage more research on horticulture as a part of the overall cropping system, particularly in accessible areas where production of fruits and vegetables could provide considerable income. The success of the recent changes in the research and extension system depends on the ability of the system to exchange information with the farmers. This means that in order to produce information useful to poor farmers, extension and research staff will have to be trained to communicate with poor farmers in general, and particularly with female and low caste farmers.

4.32   While these steps will be effective in reaching the moderately poor, it has to be recognized that many of the poor have too little land for it to be efficient for general agricultural services to focus on them - such general programs should instead concentrate mostly on the non-poor and upper ranges of the poor, where returns will be greater.

---

[7] See, for example, B. Martens; *Economic Development That Lasts - Labour Intensive Irrigation in Nepal*; ILO; 1989.

Poverty and the Environment

4.33   Poverty is both a cause and effect of environmental degradation. Lacking assets and choices, open access to Nepal's natural resources has provided the poor with the only feasible short-term mechanism for their survival, albeit at declining rates in both the Hills and Terai. At the same time, at a micro-level, environmental degradation imposes significant direct costs on them. The most important of these are low and declining farm productivity, and rising production and household maintenance costs.

4.34   Nepal's agriculture suffers from a chronic nutrient deficit. In the hills this is the direct result of the human and livestock populations overtaking the productivity of forests and pastures. The average hill family maintains four cattle and two buffalo, and requires about 3.5 ha. of uncultivated forest or pasture land to sustain each hectare of land under cultivation. Fodder shortages translate directly into a lack of manure, declining soil fertility and declining crop yields. Restoring fertility will have a major effect on crop productivity in the hills, although for the poorest families this will have only a limited effect on household incomes because their holdings are so small.

4.35   Increasing production costs stemming from environmental degradation take several forms. Most important for the poor are the increasing costs of fodder and fuelwood provision borne by family labor, women and children in particular. As forests and pasture become degraded, fodder and fuelwood must be fetched from increasing distances and from less productive sources, using time that could otherwise be spent in income-earning activity, or in the case of children, in education. As a further result of forest depletion, dung is increasingly burnt as fuel instead of being returned as fertilizer to the soil. It is estimated that about 8 million tons of dung are burnt each year, equivalent to one million tons of foregone grain production. Alternative fuels, such as kerosene, electricity and biogas will generally be beyond the financial reach of the poor.

4.36   The high Himalaya is an extremely young and unstable area. Massive erosion is a natural phenomenon under these conditions, and the macro effects of erosion and hydrology - both within the hills, in the Terai, and beyond - would occur regardless of the actions of man. In these circumstances it is important to distinguish carefully between that which is inevitable and that which is potentially controllable. Those effects directly attributable to human activity include deforestation; loss of soil productivity by over-use; and increased erosion due to extension of agriculture and roads onto fragile slopes. Many of these are a consequence of over-population, rather than environmental mis-management as such. The solutions include: (i) curbing population growth; (ii) greater user management of resources (especially forests); (iii) increasing the supply of inorganic fertilizer; and (iv) improving the administrative capacity to manage public resources. The steps required have been outlined elsewhere[8] and many are already being pursued by HMG (eg. the formation of forestry user groups). The poor will, in general, be least able to protect themselves from the consequences of environmental degradation, and lacking

---

[8] See for example the ERL report cited earlier.

alternatives, will be forced to contribute to a worsening situation. It is also important to be aware of the costs to farmers of environmental measures (eg. erosion control, medium term abstinence from using forests while they regenerate, etc.), and to be prepared to compensate them if necessary.

B. Off-Farm Employment and Incomes

The Formal Sector

Sectoral Employment and Issues

4.37   The formal sector (including tourism and the construction industry) contributes about 25% of GDP, and employs approximately 10% of the labour force.[9/] It is still in the very early stages of development; is concentrated almost exclusively in urban areas, and exists in virtual isolation from the rural subsistence economy. For most Nepalese - and almost all the poor, their involvement with the formal sector extends only as far as infrequent bus trips or kerosene purchases, and perhaps construction employment or contact with Government services on an exceptional basis.

Manufacturing

4.38   The manufacturing sector employs about 150,000 persons - mostly in small grain mills and brick works. Impressive rates of employment growth have been achieved over the last ten years (10% p.a.), but this has been on a small base, and in absolute terms is still insignificant. Manufacturing has been adding jobs at the rate of 9,000 per year - compared with labour force growth of about 250,000 p.a..

4.39   It is unlikely that manufacturing jobs absorb the poor - because only about 20% of jobs are unskilled, and because plants are located in urban areas where the poor are not. (Although the moderately poor are probably absorbed in textiles and brickmaking.) The wages paid for unskilled work are slightly, but not much, higher than agricultural wage rates (in the neighborhood of Rs. 25-50 per day). Employment at these wage levels would not be sufficient to generate above-poverty household incomes, unless all adult family members were employed.

4.40   Industrial expansion is hindered by the factors described earlier (poor access, high costs, and the effects of India's trade regime). To the extent that rapid growth can be sustained, employment growth will probably be slower, because the labour-output ratio is falling. This is consistent with the experience in other developing countries, where industrial growth has resulted in less employment generation than had been hoped for.

[9/] Note that except for government and manufacturing activities data on the formal sector is very poor, and estimates throughout should be treated with caution.

Construction

4.41   Construction is the single most important source of off-farm employment of the poor. We estimate it provides about 350,000 jobs[10/] - perhaps two-thirds of them unskilled. These jobs almost certainly employ the poor - the wage levels are low enough to be self-targeting, and in many cases they are located in areas where the poor live.

4.42   Typically only 30% of construction expenditure translates into wage incomes - due to high materials transport costs, high contractors margins (often 30-40% of costs) and the use of urban-based contractors which results in as little as 20% of expenditure remaining in the project area. There is scope for increasing the employment-generation effect of expenditure on public works, by adjusting contractual arrangements (eg. using local petty contractors), by shifting in some instances to more labour-intensive methods, and when feasible, selecting smaller rural works. Various programs have had success in raising the proportion accruing to labour to 50% without raising costs - thus creating two-thirds more employment at the same level of construction expenditure. One promising area is in increased maintenance of roads - existing roads are often under-maintained, and it is estimated that 80% of expenditure translates into local labour incomes.

4.43   Public works and construction are not, in the long run, going to be a major solution to poverty in Nepal or anywhere else. However:

- intensified expenditure on rural works *can* be an important stop-gap measure to support the rural poor until population programs and agricultural improvements take effect;

- some of the adjustments outlined above can increase the poverty alleviation impact per unit expenditure;

- these expenditures also create productive infrastructure in rural areas which, if there is careful project selection, will also benefit the rural poor; and,

- there is room for a substantial increase in expenditure on maintenance, which could have a particularly beneficial impact on the rural poor.

4.44   There are clearly trade-offs involved in maximizing construction employment. *It is recommended* that HMG undertake analytical work to investigate the scope for increasing the labour-intensity of public works at acceptable cost, to frame a program for donor support. This should include (i) an analysis of existing labour-intensive initiatives in Nepal and elsewhere; (ii) an assessment of efficiency trade-offs; (iii) an examination of contractual arrangements and employment conditions; and (iv) criteria for maximizing employment plus project benefits.

[10/] Full-time equivalent jobs. Data on the construction industry is sketchy, total employment may be lower than these estimates.

Hydroelectricity

4.45   Hydroelectricity is one of the few obvious natural resources which Nepal can potentially exploit to generate growth. Hydro exports to India could generate in the order of US$300 million equivalent in annual revenue within the next 20 years.11/ This represents a 10% increase in GDP, potentially an important source of financing for investments and services which could reduce poverty.

4.46   Government should be all means proceed with the development of hydro power both for export and for the beneficial effects of rural electrification, however some cautions are in order. There is very little direct employment benefit from hydro. The benefits thus depend on (a) the distributional effect of added foreign exchange earnings, and (b) the propensity of Government to spend on services which benefit the poor. The impact of resource-based expansion in other countries (especially oil-revenues)12/ has been a currency appreciation which shifts the terms of trade away from tradeables and in favor of non-tradeables. Thus agriculture, manufacturing, and tourism lose, and the non-traded sectors (eg. especially real estate) gain. It is thus possible that without strong and effective counter-measures hydro development may not benefit the poor - and could possibly make them worse off. Finally, care should be taken that during the development phase resources are not diverted from other priority sectors.

Government, Trade, and Services

4.47   The public sector (including parastatals) employs about 130,000, 13/ By the nature of its activities the Government tends not to employ the poor or uneducated, although there are about 25,000 peon-level jobs which may employ the medium-poor. The capacity to expand public employment is limited, at any rate, by the rate of growth of government revenue.

4.48   The main effect of Government expenditure on the poor may be through increased economic activity in small towns and road-head bazaars, where the salaries of growing numbers of public servants have fueled phenomenal local growth. It is not clear, however, that the poor have benefited. It is claimed that the businesses, teashops, and services catering to the public servants are not operated by the poor, and purchase their inputs largely from formal sector sources and Kathmandu. When the mission met with poor farmers living within a couple of kilometers of such a road-head bazaar, it appeared that this transformation had had no effect on them whatsoever.

4.49   There is no reliable data on formal sector employment in trade and services (eg. large retail and wholesale firms, financial and consulting services, trucking and airline companies). We estimate it to be about 180,000 (with another 500,000 in self-employment and the informal sector -

---

11/ Assuming export of 1,000 MW at a 50% capacity factor, sold at Indian rupees 0.90 per kwh (US$0.07/kwh).

12/ See A. Gelb; Oil Windfalls, Blessing or Curse?; 1989.

13/ Data is incomplete, and may understate total public sector employment.

see following section). It is unlikely that these firms employ the poor - for the same reasons that the government and industrial plants do not. It needs to be recognized that any formal sector job - even at low wage levels - is keenly sought after for the security of cash income it provides. Access is heavily influenced by connections of caste and family, and the ability to pay bribes, making it unlikely that the poor will be hired.

Tourism

4.50   Although tourism is often cited as a potentially important source of incomes - especially in the hills, total tourist receipts contribute only 2% of GDP, and the effect on personal incomes is limited. A large proportion of tourist expenditures go on imports and the services of urban-based firms. ERL estimates that only 6% was spent in rural areas in 1988. 75% of tourists never leave Kathmandu or Pokhara, and when they do they only spend an average of $3-4 equivalent per day in villages. Even this tends to be concentrated along two or three narrow trekking routes, and among very specific groups (eg. Sherpa porters). While the effect on individual villages is of course great, the impact on poverty alleviation in aggregate is very small. We estimate that tourism employs about 10,000 persons in the formal sector (eg. hotels, tour companies, etc.), and generates another 10,000 full-time equivalent jobs in rural areas.

Potential Contribution

4.51   Table 4.5 illustrates the importance of formal sector employment, and the potential level of employment under moderately optimistic assumptions (3.5% p.a. sustained GDP growth, and 6-7% p.a. growth in manufacturing and services).

Table 4.5: Current and Projected Formal Sector Employment

Sector	Estimated 1990 Employment	Employment Growth Rate	Projected 2010	Annual New Jobs (in 2010)
Manufacturing	150,000	6.5%	528,500	34,400
Construction	350,000	3.0%	631,000	19,000
Tourism	10,000	6.5%	35,000	2,300
Public Sector	130,000	3.0%	235,000	7,000
Trade	60,000	6.5%	211,000	13,700
Transport	18,000	3.5%	36,000	1,250
Services	100,000	7.0%	387,000	27,100
Total Formal Sector: with Construction	818,000		2,063,500	104,750
without Construction	468,000		1,432,000	85,750

4.52   The formal sector might be expected to provide about 2.1 million jobs by 2010 (1.4 million excluding construction) - employing 15% of the expected labour force. New employment growth, if sustained at the relatively high proportional levels, could absorb perhaps a quarter of the marginal labour force growth of 380,000 p.a., a useful, but not dramatic contribution to absorption of the poor in off-farm activities. Even this would involve adding jobs at about two-and-a-half times the current rate. The implication is that while HMG should certainly continue to pursue measures which improve the environment for formal sector growth, the most important efforts to raise incomes will have to be focussed elsewhere.

The Informal Sector

4.53   Informal sector activities are the principal occupation for very few households - although they contribute small amounts to the incomes of many. This is because most households still have some landholdings, and because opportunities have been limited by the virtual absence of a cash economy. On the basis of MPHBS data it is estimated that some 11% of the rural labour force are engaged in off-farm activities in the non-formal sector as their primary occupation, and 32% of the urban labour force. This represents some 1.2 million persons, broken down as shown in Table 4.6.

Table 4.6: Estimated Informal Sector Employment - 1990
(By main occupation in thousands of persons)

	Rural	Urban	Total	Proportion of Informal Sector
Sales	231	78	308	26%
Service	121	57	178	15%
Production	254	88	343	29%
Transport	12	18	29	2%
General Labourers	285	51	336	28%
Total	903	291	1,194	100%
Proportion of Regional Labour Force	11%	32%	13%	

Source: Mission estimates based on MPHBS data.

Many rural households are engaged in multiple occupations on a part-time basis - correcting for this implies that about 15% of the LF may be engaged in informal sector activities, broadly defined (1.4 million full-time equivalent jobs).

4.54   There is substantial household <u>cottage industry</u> production in rural areas, but most is not marketed. Surveys indicate that probably only 20% of output is sold, and that cottage industry and trade accounts for only 4% of incomes among sample households (and less among the poor).[14] With respect to more structured small industries, it is estimated that some 3,500 cottage and small industry (CSI) units have opened over the last five years, but

---

[14] See Chapter 2.

many fail and there is no reliable estimate of how many are currently producing, or the numbers they employ. They tend to be owned and operated by the non-poor, who can take advantage of subsidy and incentive programs. Many utilize only family labour, although some larger ones (especially in urban areas) do generate employment for the poor.

4.55   Small businesses consist mostly of small shops, tea stalls, eating houses and lodges. They generally provide a supplementary source of income for agricultural families, and are seldom profitable enough to sustain a household on their own. There is no data on the scale of such businesses. They have grown very rapidly along roads, and anecdotal evidence suggests they are also failing very rapidly as a result of over-supply. In urban areas many of the poor are absorbed in hawking and street vending (mostly textiles and vegetables) and services (eg. shoe repair, hair-cutting, etc.).

4.56   Portering provides incremental cash income for many in the hills, although usually only for a few days a month. (The vast majority of the loads in the hills are self-carried). It pays better than agricultural work, but is strenuous and intermittent, and involves significant employment costs (in terms of waiting and meals). Expansion of the road network has, if anything, increased portering opportunities, as more goods are distributed from roadheads.

4.57   Occupational castes include blacksmiths, tailors, cobblers, metalworkers, etc. Census data and surveys show a dramatic decline in the numbers engaged in these activities. The traditional patron-client relations which supported them have broken down, and demand for their products has been eroded by manufactured goods. Some have successfully made the transition to urban craftsmen or vendors, but most have become landless labourers. They face particular difficulties because, as untouchables, many forms of employment are not in practice open to them.

Informal Sector Incomes

4.58   Marketed non-formal activity contributes relatively little to incomes, especially among the poor, although there is substantial production for home consumption (eg. household food processing - which constitutes the largest share).

4.59   We estimate that roughly 18% of total household income may be generated by non-formal sector activities (up to 34% in urban areas) - although of course there is wide variation between villages and households. Average non-formal sector incomes are estimated (very roughly) at about Rs. 250 monthly per household in rural areas (US$1.62 equivalent per capita), and at Rs. 850 in urban areas (US$6.20 per capita).15/ There is of course a marked discontinuity in returns between better off families who run small businesses, and returns to the subsistence non-formal activities in which the poor usually engage. We have seen (Chapter 2) that daily returns to off-farm labour are both lower than those to on-farm labour, as well as being much lower among the poor than the non-poor.

15/ 1984/85, expressed in 1988/89 prices.

4.60   The incidence of poverty among informal sector workers follows much the same pattern as for the population as a whole, except in the urban hills, where 73% are below the poverty line (as opposed to 15% among the general population) - highlighting the marginal existence of those in activities like street vending, carpet and handloom production, and laundry and other services for which particularly high levels of poverty are reported.

Dynamics and Potential Contribution

4.61   The informal sector in Nepal is in transition - traditional off-farm activities are dying out, and are being replaced by more cash-economy-oriented opportunities in urban areas and along roadheads. Informal activities are still seen by the majority as incremental sources of income - the transition to conceiving of them as potential principal occupations is only taking place slowly. Given the underdeveloped state of the informal sector (compared for instance to other countries in Asia and Latin America), there is still substantial scope for expansion. In aggregate, of course, such expansion can at best follow demand led by growth in agriculture and the formal sector. However, there is a backlog of untapped potential due to the very low level of monetization and access in the past - there thus is scope for a catch-up effect which may allow informal sector growth to exceed growth of the economy as a whole for some time to come. In addition, the existence of isolated hill micro-economies also creates opportunities, (eg. for food processing, for localized import-substitution, and for production by poor households to supplement their own consumption).

4.62   There are a number of constraints - none of them overwhelming.

(i)   The most confining is probably locational. Non-formal sector opportunities for the most part do not exist where the poor currently are (i.e. in areas with poor access and limited cash economies) - to take advantage of them they will have to migrate. Although at the expected high rates of urban LF growth migration in pursuit of non-formal opportunities represents an opportunity for some, not a solution for many.

(ii)   It is reported that lack of information about markets and techniques (more than lack of marketing networks) often prevents the poor from taking advantage of opportunities.

(iii)   Nepal does not have the sort of regulatory framework that constrains non-formal sector activity in other countries - although the power of local officials and police reportedly results in a lot of unofficial taxation of small enterprises.

(iv)   There are few explicit barriers to entry - rather the non-poor will be better positioned (because of their asset base, education, and connections) to take advantage of commercial opportunities - the poor will almost inevitably be limited to the more marginal informal activities.

4.63   One can identify a number of potential areas of opportunity mostly in retailing, tourism, food processing, service and repair works. On the basis of fairly conservative assumptions, it looks like the informal sector could create about 3 million jobs by the year 2010, accounting for 17% of the labour force. (see Annex II.5).

4.64   The policy measures available to facilitate non-formal sector growth are limited. Perhaps the most effective will be to open up access through road building. Basic education can also equip the poor to better take advantage of opportunities as they arise. The effectiveness of credit and entrepreneurship interventions is questionable - there is little evidence that these are binding constraints, and experience with the CSI project suggests that they: (i) will be taken advantage of by the non-poor; and, (ii) will result in few long-term productive investments which would not otherwise take place. Programs with a very small-scale focus (eg. SFDP) have had some success in getting the poor involved in incremental non-formal activities (eg. purchasing a goat, poultry raising), mostly through group formation and intensive support; although their effectiveness in raising incomes has been mixed.

V. THE SOCIAL DIMENSIONS OF POVERTY

A. Nutrition and Access to Food

5.1   Thirty-six percent of the population of Nepal consumes less than the estimated minimum calorie requirements. Nationally, more than 50% of children under six are stunted,1/ with local studies finding rates as high as 80%. These rates of malnutrition are as high as anywhere in the world and are comparable to those found in the African Sahel during periods of drought. Food security is complicated by high levels of disease and parasitic infestations so that consumption of adequate calories is a necessary, but not sufficient, condition to prevent malnutrition.

5.2   Access to food is not, of course, necessarily linked to local food self-sufficiency; in Nepal however the two are closely tied. Firstly, because in many areas the cost of bringing in foodstuffs is sufficiently high that they will not be affordable at any expected income levels in the near future. Secondly, production of foodgrains for own consumption constitutes the largest share of income among the poor, so in the absence of substantial off-farm employment opportunities, raising incomes (the correct solution to food insecurity) will consist largely of raising foodgrain production.

Nutrition

5.3   The last national nutrition survey in 1975 found 51% of children under six were stunted. Rates of stunting are somewhat higher in the hills than in the terai, but nowhere are they below 45%. More recent local surveys have found similar rates, with some areas having 80% stunting among children under five.

5.4   Inadequate access to food is clearly not the sole cause of malnutrition in Nepal. About 30% of households have inadequate access to food, while more than 50% of the children are severely malnourished. Moreover, perhaps only 10% of the households have access to so few calories that the children would be expected to be severely malnourished. Children in Nepal suffer from a combination of inadequate access to calories and very high rates of disease (the two week morbidity rate from diarrhea for children under 5 is 31%). These form a cycle in which disease leads to malnutrition and malnourished children are more susceptible to disease. In addition, most children also suffer from gastro-intestinal infections and parasites which can cause the loss of 20% or more of the calories consumed and weaken children further. Local studies show prevalence rates of 86-91% for intestinal pathogens.

5.5   Case studies comparing nutritional and economic status in Nepal show that there is some relationship, i.e. wealthier households have somewhat fewer malnourished children. But the change in economic status necessary to make a significant difference in nutritional status is quite

---

1/ Stunting: less than 90% of the National Academy of Science median height-for-age. Stunting at this level represents a serious health risk and is associated with increased rates of morbidity and mortality.

large: with 50-100% increases in income needed before measurable declines in stunting are observed (see Annex II.8). It would be unrealistic to expect that income generating efforts alone could produce a change in income large enough to affect nutritional status for many of the poor.

5.6   Reducing malnutrition requires insuring access to sufficient calories and reducing the prevalence of parasites and disease. Specific interventions are possible to improve hygiene, promote appropriate weaning practices, manage disease episodes more effectively, improve sanitation and water supply and promote breastfeeding. In addition, effective health programs such as immunization will help break the disease-malnutrition cycle (see subsequent discussion of health and the poor).

Adequacy of Food Supplies

5.7   The percentage of households consuming less than recommended levels of food is highest in the rural hills, 47%, followed by urban areas, about 40%, and the mountains, 31%. 2/ Even in the rural terai, which is a food surplus producing region, 23% of households do not consume enough food to meet the requirements.3/

5.8   The proportion of income spent on food is a good indicator of poverty and of food insecurity. Typically, at very low levels of income, expenditures on food remain constant as a proportion of income until caloric requirements are met, or are close to being met. Then as income increases, the proportion of income spent on food declines. This is true in Nepal, but significant declines in the proportion of expenditure on food occur only at higher income deciles.

5.9   In rural areas, food expenditures remain over 60% of total 4/ expenditures through the 6th or 7th income decile, suggesting that until such levels of expenditure, households do not feel they are meeting their food consumption needs. It appears that a per capita expenditure level of roughly Rs. 200/month (1984/85 prices) is needed before households greatly decrease the proportion of expenditure on food. Only about 35% of the households in rural areas have this level of income. Expenditure patterns for most of the population (see Chapter 2) are similar to those of very poor people in other countries who spend high percentages of their income on food, and specifically on foods that maximize calories for each unit of money spent.

2/ The National Planning Commission calculates per capita consumption requirements of 2140 kcals/person/day in the terai and 2340 kcals/person/day in the hills and mountains, based on WHO guidelines adjusted for climate and household composition.

3/ Based on actual expenditures for food, including the value of own production consumed within the household.

4/ Total expenditure includes the imputed value of self-produced food, non-purchased fuel and the rental value of owned homes. The inclusion of the latter two makes it appear as if the poor in Nepal are spending relatively less on food than the poor in other countries where the lowest income groups often have 80% of expenditures on food.

Regional Food Balances

5.10 With few substantial sources of foreign exchange, Nepal is primarily dependent on its own agricultural production to meet consumption needs. A net exporter of food grains throughout the 1960s and 1970s, Nepal now produces less than its own consumption requirements; total current production of grains meets only 90% of requirements and is declining. While foodstuffs can in theory be imported to make up the shortfall, in fact they are not, because of limited purchasing power and high transport costs. In fact, it appears that the net flow of foodgrains is still out of Nepal to India.

5.11 Some inaccessible hill and mountain districts produce less than half of the food they need. Caloric availability resulting from regional production of grains and potatoes is shown in Table 5.1. Due to low incomes and poor transportation, local food <u>availability</u> is synonymous with local <u>production</u> in most mountain and many hill districts.

5.12 All the terai regions produce a surplus except for the eastern terai. The surplus produced in the terai does not necessarily go to deficit areas in the hills. Anecdotal evidence suggests that much of it goes to India. In the hills and mountains only the western mountains produce a surplus.

Table 5.1: Regional Per Capita Production of Grains and Potatoes
(kilocalories/person/day)

Development Region

Ecological Zone	Far West	Mid-West	West	Central	Eastern	Nepal
Mountains	848	795	2656*	930	1413	1035
Hills	846	1201	1219	1407	1741	1330
<u>Terai</u>	<u>2197*</u>	<u>2140*</u>	<u>2253*</u>	<u>2262*</u>	<u>1659</u>	<u>2046</u>
Nepal	1339	1499	1562	1795	1664	1635

* Indicates regions producing adequate or surplus food.
  Estimated minimum requirement (assuming 80% of calories provided by grains and potatoes):

                    Hills and Mountains: 1872
                    Terai   : 1712

Source: DFAMS and CBS.

5.13 Population growth has been outstripping food production in most areas of the country and most importantly in the terai, the major surplus area (Table 5.2). Given the dependence of incomes on agriculture, the decline in per capita production of calories suggests a similar decline in per capita income and a decrease in food security. While population growth **may** not continue at the same high levels, it is unlikely that food production can continue to increase in the same manner as it has done over the last two decades - through expansion onto marginal lands (see para. 5.17).

Table 5.2: Population Growth Rates (1971-1981) and Growth Rates of Agricultural Production (1967/68-1987/88)

Annual Growth Rate of:	Mountains	Hills	Terai	Nepal
Population	1.4%	1.7%	4.1%	2.6%
Agriculture	0.2%	1.4%	1.5%	1.4%

Source: DFAMS and CBS.

5.14 In addition to regional food shortages, there are wide seasonal variations in food supply. In the absence of reliable water supplies, foodgrain production varies by as much 40% from year to year. Thus every four or five years there are major seasonal shortages (see Figure 5) - which can be catastrophic for the poor.

Figure 5.
Annual and Trend Variation in Food Production
(Kcal/person/day from production of cereal and potatoes)

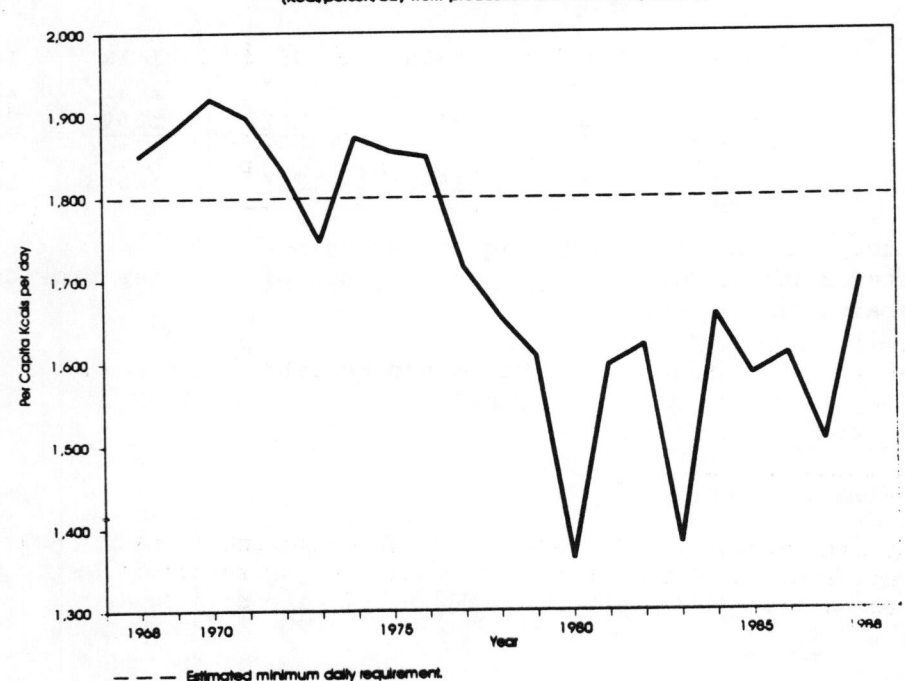

--- Estimated minimum daily requirement.

Food Prices

5.15   The rural poor in Nepal produce only about 65% of the food they consume, so food prices have a significant impact on their food security. In the inaccessible hills, local supply and demand largely determine prices because transport costs make imported foods prohibitively expensive. For example, transportation costs for grain from the terai to a hill town 2 days from the roadhead doubles the price. In the terai, prices are largely determined by the Indian market. This dominance of the terai markets by Indian markets and the inaccessibility of the hill markets means there is very little scope for government price policy to improve food security.

5.16   Inaccessibility and transport difficulties also exacerbate seasonal price fluctuations. In remote districts, fluctuations may be 40-60% compared to a national average of 20-25%. The poor are hurt twice by the wide price differences because, in order to pay debts, they are often forced to sell crops immediately after harvest when prices are low, and buy food during the dry season when prices are high.

Food Balance Projections

5.17   Projections show that with current agricultural growth rates, only a drastic reduction in population growth will improve per capita food production (column 1 in Table 5.3) and this is unlikely to occur. Improved agricultural performance (column 3) and moderate decreases in population growth will improve per capita food production somewhat, but domestic food supplies will remain below levels necessary to meet minimum needs. All these projections are dependent on continuing increases in cropped area and cropping intensity which are not sustainable at current rates. If, as is likely, limits to the expansion of farming land are reached before 2010, then in the absence of major yield improvements sharp decreases in per capita production will occur (column 4). 5/

Table 5.3: Projected Daily Per Capita Availability of Calories from Grains and Potatoes
(Kcals/person/day)

Agricultural Growth and Population Projection

	Current Ag. Growth and Low Pop. (1)	Current Ag. Growth and Medium Pop. (2)	Improved Agriculture & Medium Pop. (3)	Restricted Land Expansion & Medium Pop. (4)
1990	1,499	1,499	1,499	1,499
2010	1,621	1,485	1,730	1,211
Corresponding Grain Deficit in 2010: (000 mt.)	-567	-1,040	-232	-1,950

Source: Annex II.8.

5/ Long-term agricultural productivity is dependent on maintenance of some forest land for fodder and fuel resources, so reserving land for forest is economically rational although politically difficult.

Improving Access to Food

5.18   For the foreseeable future it appears that only the agriculture sector has the potential for substantially improving incomes among the poor, both directly through increased production and indirectly through increased demand for labor and decreased food prices (see Chapter 4). Until the transportation system improves dramatically, producing food locally will be much more efficient than importing food, even though local production costs are often higher and yields lower. In addition, local production largely determines local demand for labor and therefore local wages.

5.19   After improvements in agricultural productivity, improvements in infrastructure may have the most significant long-term impact on food security through decreased prices for food and agricultural inputs. Comparison of prices in adjacent hill districts show that districts without motorable road linkages to the terai have food prices 50-100% higher than districts with such road linkages.

5.20   A third method of improving long-term food security involves improvements in the utilization of current household resources. Fifteen to twenty percent of the population has access to more than the minimum recommended calorie levels, but is malnourished nevertheless due to disease and parasites. Improved health services, focusing on prevention - better water supplies, hygiene, improved feeding practices, could all have a substantial impact on the number of people malnourished.

5.21   Improved food storage facilities combined with a credit program could also potentially improve utilization of current household resources by allowing poor families to hold crops until prices were higher rather than selling them immediately after the harvest. Poor households would benefit through lower post-harvest storage losses, higher prices for the crops they sell and/or less need to buy food when prices are high.

5.22   The programs most frequently associated specifically with food security in Nepal are food subsidy and distribution programs. These programs provide what amount to short-term increases in incomes through lower food costs or free food. These and other food security programs are discussed in more detail in Chapter 6.

B. Education and the Poor

5.23   As was pointed out in Chapter 2, the poor participate *much* less in education than do the non-poor. Official enrollment rates have grown from about 22% in 1971 to about 85% today.[6] But this growth has been fueled largely by increases among the non-poor, enrollment by the poor - and especially poor girls - is substantially lower.

---

[6] Gross primary school enrollments according to Ministry of Education statistics - actual rates are probably substantially lower.

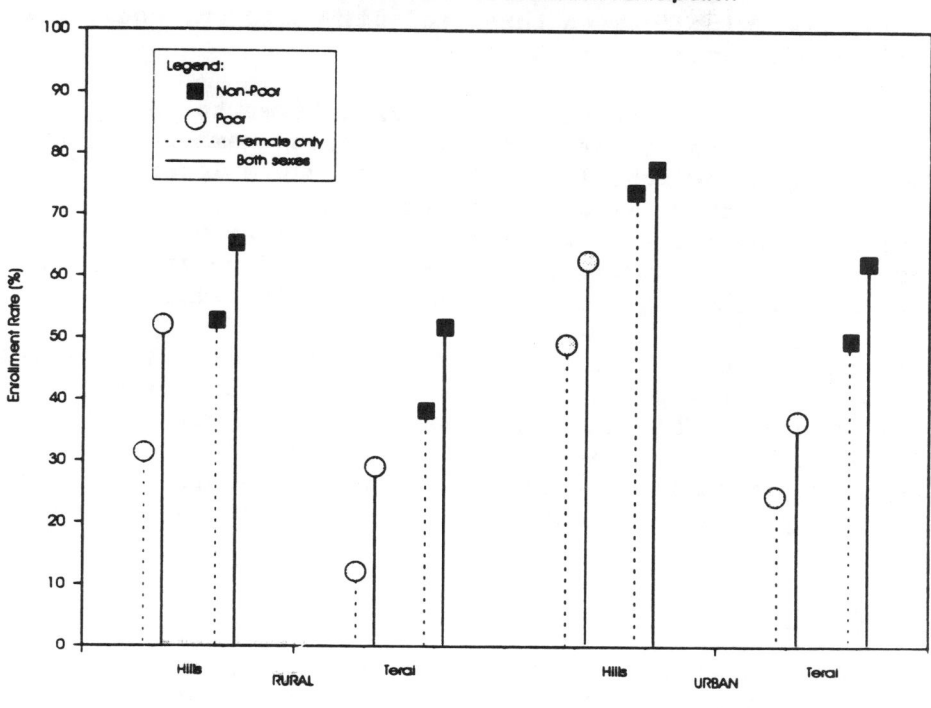

Figure 6.
Income, Gender and Education Participation

Legend:
- ■ Non-Poor
- ○ Poor
- .... Female only
- ——— Both sexes

MPHBS data show enrollment rates among the poor of about 30% in the terai (compared to 50% among the non-poor), and 55% in the hills (73% among the non-poor). At secondary levels and among girls the differentials are even more marked (see Annex II.2).

5.24   Access by the poor is hindered because they are less able to pay the direct costs of education; because they are less able to bear the indirect costs of labour forgone; and because they tend (at the margin) to live in areas which are less well served with schools.

5.25   Children's labour time is a significant source of family income - LF participation rates are higher among 10-14 year olds than for the labour force as a whole. They also provide a large proportion of domestic labour - tending livestock, caring for siblings, cooking, and carrying water and fuel - freeing older family members to engage in essential off-farm employment. Time-use data shows that girls above the age of 10 are engaged almost full-time in such pursuits, and about half time (3-4 hours per day) below the age of 10. There are marked differences between poor and non-poor female children's work hours in the terai - less so in the hills (perhaps due to the universal work burden imposed by poor access). Children also play a key role in meeting peak agricultural demand during harvesting and transplanting - at which times the family cannot free them for school.

5.26   The direct costs of school attendance may be in the range of Rs. 10-20 per month (much higher in secondary school)[7] - relative to discretionary household cash incomes of Rs. 100-140 per month among the poor. Such families, with between three and six children, would likely be able to afford to send only one to school.

5.27   There are other, less tangible, disincentives to educational participation which relate to caste and ethnic group membership, and the subtle perception that the school is not a place where lower social groups are welcome. Finally, there is a perception among the poor that education is unlikely to make any material difference to their economic standing. This is based in part on accurate perceptions of limited income-earning opportunities, and of the rigid caste structure, which limits mobility.

5.28   Furthermore, some characteristics inherent in the school system discourage participation - particularly by the poor. The primary curriculum is excessively complex. Children are expected to join with a basic knowledge of the alphabet - an infeasible proposition in the case of the poor who come mostly from illiterate households. Similarly, education is provided exclusively in Nepali, which is less likely to be the mother tongue of the poor.

The Education System

5.29   The education system as a whole suffers from poor facilities, inadequate teacher training, a curriculum that is too academic and often irrelevant, and excessive reliance on rote learning and examinations. Coverage, although much improved, remains insufficient, and many children are still an hour or more's walk from school. The internal efficiency of the system is very low (taking an average of 13 person-years of instruction to produce an elementary graduate), due to very high drop-out and repetition rates.[8] There is thus substantial scope for improved effectiveness within the existing system. However the two overriding problems are growing financial constraints and non-attendance of children (particularly girls) who are within range of a school.

5.30   Expenditure on education is about 1.9% of GDP - currently equivalent to US$3.10 per capita. This puts Nepal among the lowest spenders on education in the world, both as a proportion of GDP, and in absolute terms. In aggregate it is clearly insufficient, since it only covers the costs of the 65% of (primary school-aged) children who are in in primary school, and the 20% of those (secondary-aged) who are in secondary school. The allocation of funds could be improved (for example the university has 2% of the students but 23% of the budget) - but this would not obviate the need

---

[7]   While primary schooling is in theory free, costs are incurred on supplies, examination fees, and often other charges levied to support school budgets.

[8]   For example 62% of those who start the first grade do not pass to the second grade.

for major additional financing if educational capacity (even at the primary level) is to catch up with potential demand.

5.31   Attempts at alternative financing arrangements have had undesirable equity effects. HMG has consciously under-funded secondary expansion, providing only about 50% of total costs, and passing the rest on to the community. As a consequence there has been rapid growth of private secondary schools - which tend to serve only urban areas and the relatively well off. The danger is entrenchment of a two-tier system, which provides highly unequal access to, and quality of, education. Nor are user fees a very useful solution in the Nepalese context. Given the high value of social benefits resulting from primary education, it is doubtful that one would want to use financial instruments which would further discourage participation. Furthermore, with the very low level of cash incomes among the poor, there is likely to be a serious ability-to-pay constraint.

5.32   Insufficient capacity may not, however, be the binding constraint. Many children who live within range of a school are not attending. The reasons for non-attendance are complex, and include:

- other demands on the use of childrens' time;

- the low private returns of providing education (especially to girls, who are lost to the family at marriage),

- inability to pay the private costs of education,

- cultural factors, and,

- perceived irrelevance and poor quality.

The determinants of school attendance are not well understood, and require more analytical work to determine what the most cost-effective interventions may be to get children into school. However the time-use data suggest that increasing enrollments (which amounts largely to increasing female enrollments) may depend on designing measures which reduce the work burden on girls.

<u>The Role of Education in Poverty Alleviation</u>

5.33   What then is the relative importance of education as a poverty alleviation strategy in Nepal - both for the country as a whole (i.e. in contributing to economic growth in general), and for poor households specifically? The corollary to this question is: What is the optimal level (and type) of education to provide to have an impact on the incomes of the poor?

5.34   There has been substantial research on the productivity effects of education in developing countries. Most indicates that agricultural output increases (by about 10%) as the result of primary education, that the main impact is on the use of modern inputs, that the impact is higher in lower technology/poorer environment conditions, and that basic education is generally adequate for traditional farming systems.

5.35   Studies in Nepal estimate the increase in farm output as the result of primary education at between 10% and 20% (the latter in conjunction with complementary inputs),9/ with higher gains on modernizing farms than on traditional ones. Jamison and Moock10/ found education effects significant only in the case of wheat (an introduced crop), confirming other studies which found education to be most important when adapting to changed circumstances. Two studies also found definite evidence of a threshold level of education of about six or seven years - below which the impact on incomes was small. (This may however reflect the inefficiency of the education provided - requiring seven years to produce a primary education - rather than implying a need for lower secondary education to achieve an income effect).

5.36   The effectiveness of basic education in raising _personal_ incomes thus appears to depend on having a constellation of supporting factors - perhaps good land, an accessible location, access to modern inputs - which are unlikely to be available to the poor.

5.37   The poor may in fact be following a rational strategy in not sending their children to school - if there is an absence of off-farm employment opportunities, if the perceived relevance of both the curriculum and quality is low, or if the opportunity cost in labour forgone is higher than the returns to family income (or all of these). And _especially_ if they are correct in their perception that a fairly high level of education (eg. secondary or above - which they have little hope of achieving) is needed to make a marked difference in expected incomes.

5.38   On the other hand, no country has made the transformation to the sort of growth path which is necessary to eliminate poverty in Nepal without widespread general education to at least the primary level. Analysis by USAID suggests that providing universal primary education in Nepal would raise aggregate agricultural output by about 6% - paying for itself in the first year. Furthermore, there are a number of indirect but very important effects, including:

- the overwhelming impact of female education on fertility;

- the effect of general (and especially female) education on hygiene behavior (and thus on nutritional levels);

- the role female education plays in breaking down the traditional cultural seclusion of women, leading to the higher female LF participation rates needed by the poor (especially in the terai); and,

---

9/ See Annex III for references on the returns to education in Nepal.

10/ D. Jamison and P. Moock _Farmer Education and Farm Efficiency in Nepal_, The World Bank, 1984.

- the fact that education of the poor builds their confidence, improving their capacity to deal with creditors, landlords, and official services; and to form themselves into self-reliant groups.

5.39   There are thus substantial externalities involved which result in the social returns to basic education exceeding the private returns by some substantial margin. HMG should be willing to consider measures to compensate individuals to overcome this differential - by, for example, investing in improved primary schooling capacity, curriculum reform, and even scholarships in some instances (eg. to encourage female participation).

5.40   With respect to the optimal level and type of education, the evidence is unclear, but analysis in Nepal and elsewhere suggests that the rate of return to primary education is consistently higher than to higher levels (mostly as the result of lower unit costs). There is no doubt that basic literacy and numeracy alone can have a transformational effect on the poor. These can be imparted through non-formal training or about three years of primary schooling. Beyond that level, we do not know what the aggregate impact of further schooling would be on the incidence of poverty; although some analysis suggests that a level of lower secondary is required to make a substantial difference to personal incomes and fertility behaviour.

5.41   Higher level (eg. secondary) education can have a substantial benefit for the poor <u>individuals</u> who receive it, but is unlikely to have much effect on an aggregate scale in the absence of a transformation of the economy that would create substantial formal sector employment and/or the conditions for modern agriculture. However in the long-term, given the limited resource base, any development strategy for Nepal must rely on developing a skilled workforce.

5.42   With respect to other training and skills acquisition programs designed specifically to raise the incomes of the poor, the experience has not been encouraging (see Chapter 6 on poverty programs). While developing a skilled workforce is an important element of overall development, our analysis of non-formal sector employment suggests that the binding constraints on the poor are often not skills (which at the low levels involved could probably be most cost-effectively acquired informally or on the job), but rather access to markets and inputs, and the absence of a cash economy.

<u>Strategies</u>

5.43   HMG should continue to upgrade general education for the beneficial impact it can have on aggregate economic growth. However to the extent that the objective is poverty alleviation, it should concentrate its efforts on:

- literacy and numeracy,

- lower-to-middle level primary education, and,

- education of girls - for the effect this can have on fertility, hygiene, and nutrition.

A package of specific measures should include those outlined below.

5.44 <u>Allocate Sufficient Resources to Primary Education</u> - if Nepal were to achieve universal primary education by the end of the century, it would require a sustained level of expenditure of about US$50 million equivalent per year (compared to a current level of $24 million equivalent). There is room for cost-effectiveness improvements (by getting more students into existing schools, and improving internal efficiency - by for example reducing repetition rates) - but in aggregate these would not substantially reduce the magnitude of the requirements, which are driven mostly by the growing numbers of school-aged children. HMG and the aid community should be prepared to provide this level of funding.

5.45 <u>A National Literacy Campaign</u>. An existing program has developed excellent materials and approaches - but reaches only 70,000 per year out of an estimated 12 million illiterates. It could be expanded to a national campaign at modest incremental cost.

5.46 <u>Measures to Improve Participation by the Poor</u>. Design of these measures - which may rely on non-educational interventions - will have to await deeper analysis of the determinants of school participation. Some measures need to be undertaken anyway to improve the utility and attractiveness, of primary education, including:

- improving the relevance of the curriculum;

- shifting away from rote learning, and reducing reliance on examinations, at least in the early years; and,

- considering automatic promotion in the first three grades (because studies show that if students stay through the first year or two they are much more likely to complete primary education.

Based on experience in other countries the following measures may also be part of the solution:

- siting schools closer to families (which allows families readier access to their children's labour);

- expanding programs which allow flexible school hours;

- recruiting teachers from the local area;

- considering instruction in languages other than Nepali; and,

- adult literacy, because literate parents are more likely to send their children to school.

5.47 <u>Measures to Improve Female Participation</u> - again, the most cost-effective solutions may involve non-educational interventions, related mostly to measures to reduce household demand for female child labour time. Two specific educational interventions would include:

(i) increasing the number of female teachers (in this regard HMG needs to postpone the requirement that all teachers hold an SLC pass - which severely limits the pool of female candidates - especially in remote areas.); and,

(ii) considering programs for female scholarships and the provision of school uniforms for girls.

C. Poverty, Population and Health

Population

5.48   There is no prospect of seriously increasing average incomes if the population continues to double every 25 years. Curbing population growth is thus the central, single most important poverty alleviation strategy for Nepal. To achieve it will require a drastic, national campaign-style solution - such as has been implemented in Indonesia or China. The current incremental approach cannot yield sufficient results to have any impact on the incidence of poverty. To achieve some stabilization of population growth (say 1.5% p.a. by 2010) will require total fertility falling to about 3 (children per woman) from its current level of 5.8 - this in turn requires a contraceptive prevalence rate of about 50%.[11]/ Only 15% of married women of reproductive age presently use any form of contraception - and most of these are accounted for by sterilizations performed after having had four to six children, and thus have little or no effect on population growth.

5.49   The Government's population efforts expanded in the early 1980's through a major sterilization program. However Nepal has not been able to develop the sort of broad-based national population program which has been successful in other countries. Essential temporary methods are delivered through a moribund health service, which lacks outreach capacity, and is staffed largely by male workers, who cannot effectively deliver family planning services to female clients. Despite a number of encouraging policy statements, empirical observations suggest that population enjoys nowhere near the level of priority that is necessary. Nepal, for example, spends about $0.10 per capita on population activities, compared to about $0.50 per capita in India and Bangladesh; similarly, the recent 8th Plan Approach paper relegated population issues to seventh place out of seven development priorities.

5.50   Nepal does not face the same cultural and religious barriers to fertility control that hinder many of its neighbors, and if the Government could put into place an effective program to promote and support temporary methods - then there is the possibility that population growth could be effectively curbed. There is some evidence that desired family size is

---

[11]/ Source: Social Sector Strategy Review; Total Fertility Rate (TFR) - is the number of children ever born per woman of reproductive age; the Contraceptive Prevalence Rate (CPR) is the percentage of married women of reproductive age practicing some form of contraception.

lower than actual family size (although perhaps not among the poor), suggesting that there is unmet demand for family planning services. Forces are thus in place for a decline in population growth rates - but only if a major national effort is made.

5.51   What is required is: (i) a major shift to <u>temporary</u> methods of contraception, and (ii) a sustained sterilization effort, but focussed on those in lower age groups who have reached their desired family size. The success of temporary methods depends on a program of sustained support at the individual level, such as can only be delivered by a cadre of well-trained (mostly female) outreach workers. Relying on incremental improvements to the current service delivery system will not yield results quickly enough to have any impact. Where such programs have worked in other countries, they have been central national priorities: well-funded, well-publicized programs that receive emphatic public support from the head of state and all levels of Government. Nepal needs to allocate major funding for such a program. Given the weakness of the service delivery system, this is admittedly a gamble, but it is one the country cannot afford not to take.

5.52   The elements of such a program might include, for example: (i) recruiting and training 25,000 highly paid female family planning workers, with adequate support and incentives to operate at the periphery; (ii) population-linked incentives in other sectors; (iii) possibly establishment of a separate delivery framework - outside of the health services; and, (iv) development of a cadre of population specialists and program managers. This would cost about Rs. 500 million per annum ($18 million - or $0.95 per capita, similar to the costs of a successful program in Indonesia). <u>It is recommended</u> that HMG develop such a program as a matter of the utmost urgency.

5.53   Provision of family planning services alone will not, of course, necessarily reduce fertility among the poor. Children have a high perceived economic value in the subsistence economy - both because they expand the family's production possibility frontier, and because they are seen as a source of security in old age. Surveys indicate that about two-thirds of families do not use contraception because they are unwilling to limit family size, largely for economic reasons.[12] This is reflected in the preference for sterilization after achieving a desired family size, and reluctance to use temporary methods. Similarly women are much more willing to consider contraception after giving birth to two or three sons - while the birth of daughters (whose economic value is lost to the family at marriage) has little impact on family planning decisions.

---

[12] UNICEF; <u>Women and Children of Nepal</u>, 1987.

5.54 The private economics of family-size decisions in Nepal are not well understood. The one available piece of analysis (of 300 terai families)13/ found only child labour and duration of marriage to be statistically significant in explaining fertility. More work needs to be done on this as part of designing major population program interventions. (It may be cost-effective, for example, to allocate resources to female employment; or irrelevant to provide family planning services if there are over-riding incentives against limiting family size.). An understanding of the household income effects of having fewer children among the poor is also important to ensure that population control initiatives do not have an adverse effect on household incomes - or if they do, to incorporate compensating measures.

5.55 Raising average incomes and increasing financial security are clearly conditions which facilitate rapid declines in fertility. However experience in other countries shows that significant fertility declines are possible in the absence of rapid economic growth. The principal determinants are average levels of education (and especially female education), the incidence of female labour force participation; and to a lesser extent levels of infant and child mortality (since total fertility may be explained in part by expectations regarding the probability of child survival.) Measures in all of these areas, in which Nepal ranks among the lowest in the world, will also pay dividends in curbing population growth.

Health

5.56 Health status is uniformly poor in Nepal - with infant mortality rates (113 per thousand) among the highest in the world, and life expectancy (52 years) among the lowest. There are remarkably high incidences of parasite infestations, diarrhoeal diseases, goitre, acute respiratory infections, and morbidity associated with pregnancy and childbirth. Cretinism, blindness, and other permanent disabilities afflict a large proportion of the population.

5.57 There is no data on health status by income level - but it is almost certainly worse among the poor than on average - because their lower nutritional status leaves them more susceptible to disease, and because they are least likely to have access to either curative or preventative health services. The health of many is sufficiently bad to impair their ability to work - draining total labour productivity, as well as reducing personal incomes and increasing the dependency burden within poor families.

5.58 The health service is at a very rudimentary level - and reaches very few - let alone the poor. There were estimated to be only 863 physicians in 1987. Health infrastructure is severely underdeveloped - in most parts it is several days' or hours' walk to the nearest health post, which is often little more than a shack lacking staff or supplies. The system is plagued by a range of institutional and service delivery weaknesses. As in most countries the composition of expenditure within health is biased (both geographically and by type of care) in favor of the

---

13/ K. Rauniyar: Demand for Children in the Nepal Terai; Winrock, 1985.

non-poor. Although there is some scope for re-allocation towards rural, primary health care, the institutional barriers to developing an effective health service in the near term are formidable (see para. 5.62).

5.59   There are, however, a number of interventions which could have a dramatic impact on the health status of the poor, at relatively little cost. These include improved nutrient retention via improved hygiene (since many illnesses are the result of poor sanitation); improved nutrition - through increased food security and better feeding practices; and measures to reduce fertility. The most useful interventions would be in the areas of female education, rural water supplies, and to some extent targeted nutrition (see nutrition section). One option may be a national hygiene campaign - which would incorporate hygiene awareness with improved water supplies and sanitation.

5.60   There are also selected <u>health care</u> interventions which are particularly cost-effective - including immunization (only 11% of children are currently fully immunized), the use of oral rehydration salts, and micro-nutrient supplementation (iron, iodine, and vitamin A). These have a disportionate impact on health status, and for each there is an existing program which is amenable to expansion, without depending on major institutional improvements. Expansion of these should be the first line of attack. Reductions in fertility will also go a long way towards improving the health status of the poor - both because health status of children increases dramatically with spacing between births, and because complications of pregnancy and childbirth account for a large share of female morbidity and mortality.

5.61   Access by the poor to health services is hindered both by cost and by the weakness of the health care system. The direct costs of medical care represent a relative small <u>average</u> proportion of expenditure by the poor (2-3% of total income, 7% of cash incomes). However numerous village-level reports suggest that it is a major cause of <u>catastrophic</u> expenditure, and is often cited as the reason for going into unsustainable debt. Unfortunately many of these payments are made for traditional cures of dubious value.

5.62   The problems of the modern health sector have been well-documented elsewhere,[14] and include:

- inadequate resources ($1.85 per capita, compared to an average of about $4.00 among low income countries);

- the absence of a sectoral strategy or expenditure plan which allocates resources where they would have the greatest health impact;

- excessive centralization of staff and resources, and unwieldy funds release mechanisms; and,

---

[14] See for example, the <u>Social Sector Strategy Review</u>.

- weak supervision and management, and a confused institutional structure.

Steps are underway to address most of these - however overcoming them will be a long, slow process involving subtle service-delivery improvements. The point is that one can improve the welfare of the poor at existing income levels by improving their health status - but it should be done through measures which are not reliant on institutional strengthening of the health service.

5.63   HMG should by all means continue with upgrading of the health service - but given the institutional barriers involved (many of which are civil service-wide), and the lack of effective access by the poor, the focus should be mostly on non-medical interventions.

VI. POVERTY RELATED PROGRAMS AND POLICIES

A. HMG's Policies and Programs

General Policies

6.1   The first few development plans during the 1960's quite correctly emphasized infrastructure investments. Recognizing that these would not create growth quickly enough the emphasis shifted (around the Fifth Plan) to raising aggregate income through production incentives and investments. Most recently the government has recognized that the benefits of this growth will reach only some segments of the population. Consequently HMG has started to concern itself with distributional aspects - particularly with assuring the provision of basic minimum needs, and with employment generation.

6.2   Historically, HMG has not in general followed redistributive policies. It uses foreign aid to finance services which benefit the poor (eg. IRDP's, health services, food aid) rather than redistribute through direct taxation. This is not an irrational strategy - as long as aid flows keep up - given the lack of resources to redistribute; although: (a) the level of taxation is remarkably low, and (b) it suggests problems of long-run sustainability of, and commitment to, these programs.

6.3   There are some notable exceptions to this trend - for example, the 1964 land reform effort, and the stated aims of the Basic Needs Program. HMG's ambivalent approach to redistributive issues has reflected the competing pressures of, on the one hand, the efficiency gains and popular support to be had from dismantling the feudal system, and on the other hand the political and economic interests of traditional elites.

6.4   Tax revenue has averaged only 6-7% of GDP, although it has risen recently (to 8.4%) as a result of measures under the structural adjustment program to improve fiscal discipline. This puts Nepal among the lowest taxing countries in the world. The ratio of revenue-to-GNP reported in the 1989 WDR is about one-third that of the average for low income countries as a whole, with lower rates reported only for Chad and Sierra Leone.

6.5   Indirect taxes account for almost 80% of tax revenue, of which import-related taxes account for half. These are progressive in that they are directly related to consumption of imported goods, which the poor do not consume. More striking is the virtual absence of income or wealth related taxes. It is admittedly difficult to administer an income tax in a subsistence economy, and a more useful proxy would be a land tax - although with very small average holding sizes, care needs to be taken to differentiate between true smallholders and surplus producers with fragmented holdings.

The Composition of Public Expenditure

6.6   The overall composition of public expenditure (Table 6.1) is relatively consistent with a poverty alleviation strategy - i.e. with the emphasis on agriculture and infrastructure (17% and 23% of expenditure respectively), and a reasonable share of public resources going to social services (16%). The problems lie more in (a) the low absolute level of financing (as a consequence largely of the low level of GDP), which results in per capita expenditure among the lowest in the world; and (b) the institutional and service delivery problems which hinder the effective use of these resources (see section on Institutional Issues below).

Table 6.1: Composition of Public Expenditure - 1988/89

	Share of Public Expenditure	Per Capita Expenditure (Rs.)	US$ Equivalent	Share of GDP
Administration	7.7%	64	$2.57	1.6%
Defense and Order	10.8%	90	$3.61	2.2%
Education	9.1%	76	$3.04	1.9%
Health	5.5%	46	$1.84	1.1%
Other Social Services	1.7%	14	$0.57	0.3%
Agriculture	17.4%	146	$5.85	3.6%
Industry and Mining	5.0%	42	$1.67	1.0%
Utilities	14.1%	118	$4.72	2.9%
Transport	9.5%	80	$3.19	1.9%
Interest, Transfer and Others	19.3%	162	$6.47	3.9%
TOTAL	100.0%	838	$33.53	20.5%

6.7   There is some room for improvement of the allocation within sectors - for example from tertiary to secondary and primary education; from curative to preventive health care; and from bulky irrigation investments to smaller schemes.[1]/ However in aggregate these will not make a substantial enough difference to the level of services delivered. That will require greater resource mobilization in total, as well as improvements in operational effectiveness.

6.8   In addition, many programs with a potential alleviation impact need a change in focus to more accurately address the needs of the poor. Many of the expenditures which are nominally justified on poverty-alleviation grounds are in fact only weakly poverty related. (See for example the discussion of subsidies and transfers, and targetted credit later in this chapter).

1/ See discussion in the _Social Sector Review_, and in relevant sectoral chapters for some specific recommendations.

The Basic Needs Program and the Eighth Plan

6.9   Recognizing that growth alone would not lift many Nepalese out of poverty in the near future, and that general development programs were not reaching the poor, HMG announced the Basic Needs Program (BNP) in 1987. The BNP was intended to achieve minimum Asian standards of living for all by the year 2000. It included the following specific elements:

Agriculture

- expansion of irrigated areas;

- intensive cultivation to reduce food deficits in hill areas (including rainfed research);

- increased use of modern inputs - through both public and private sector distribution;

- expansion of the SFDP, PCRW, and land reform, and other programs with a poverty focus.

Primary Education: Expanded teacher training; distribution of free textbooks; curriculum revision; introduction of measures to encourage girls to attend school; and expanded non-formal education.

Primary Health Care: Expansion of family planning, maternal and child health, immunization, and respiratory disease programs.

It also included measures to increase the supply of clothing and housing by the private sector.

6.10   The BNP contained most of the elements of a successful poverty alleviation program in Nepal although it suffered from some serious flaws in approach. In some sectors it was excellent (eg. education), in others it gave the correct emphasis, but did not go far enough (eg. population). The main shortcoming is that this approach is centered on intensification of existing programs - often without addressing the reasons why they have failed in the past. The target-driven nature of the program could distort efforts, and in some cases force the allocation of resources to admittedly unproductive programs in an effort to achieve unrealistic targets; there is also a danger that the input targets are not interpretted in a sufficiently flexible manner, and could stifle more imaginative interventions.

6.11   The Government realized that the BNP did not adequately address employment and income generation issues. The Approach Paper for the 8th Plan (1990-1995) emphasizes the need to generate employment, and the fact that some targetting is necessary, because general programs tend not to reach the poor. The paper proposes an emphasis on agriculture, and especially irrigation, as well as suggesting that most productive sector activities will be expanded. It does not spell out the specifics of how this is to be achieved - although this may be because it is in the nature of a broad strategy document. One area of concern is the relatively low priority accorded the population growth issue in the paper. While it is acknowledged that this is important, the document does not give the impression that policy-makers have accepted the fact that curbing population growth is <u>the</u> central element of any effort to raise incomes.

B. Poverty Alleviation Programs

6.12   The table below shows the salient features of the major poverty-related programs in Nepal. The subsequent sections summarize our assessment of their role in contributing to poverty alleviation, based on the more detailed program reviews undertaken as part of this study.

Table 6.2: Nepal - Poverty Alleviation Programs at a Glance

	Approximate Annual Cost (1988/89)	
	(Rs. millions)	(US$ million)
Subsidy and Transfer Programs		
Nepal Food Corporation	209	$8.4
Fertilizer Distribution	200	$8.0
Agricultural Credit Subsidies	48	$1.9
Irrigation Operations Subsidy	38	$1.5
Integrated Rural Development Projects	640	$25.0
Food and Feeding Programs		
Nutritious Foods Program	26	$1.0
Joint Nutrition Support Program	33	$1.3
Targetted Credit Programs		
Small Farmers Development Program	105	$4.1
Production Credit for Rural Women	12	$0.5
Intensive Banking Program	60	$2.4
Employment Programs		
Food For Work	38	$1.5
Special Public Works Program	13	$0.5
Income-Generating Programs		
Training for Rural Gainful Activities	n/a	n/a
Labour Department Centers	10	$0.4
NGO Income-Generating Programs	25 /a	$1.1 /a

---

a/   Rough estimates only.

Subsidies and Transfers

6.13   Nepal does not operate the kind of large-scale price subsidy or income-transfer programs which are common in many other countries. Those which do exist are aimed primarily at providing incentives for agricultural investments, or transport subsidies to compensate for the high cost of inputs in the hills. Expenditure on the four main subsidies programs is estimated at about Rs. 500 million p.a. (US$20 million equivalent) or only 3.3% of public expenditure.

6.14   Food Subsidies. The National Food Corporation (NFC) provides food at below market prices to civil servants, and to the population as a whole in remote areas. The principal benefit is less in the price subsidy element than in the fact that the program makes food available in areas where there is otherwise no food for sale at all for large parts of the year. However, the majority of its food goes to Kathmandu, and to civil servants to compensate for remoteness. Probably less than 25% remains as a subsidy to those living in remote areas, and while there may be some benefit to the poor, the spread of benefits in remote areas is generally at best independent of income level. (The impact of the NFC program on food security is discussed in the subsequent section on food programs).

6.15   In addition to the food distribution program, the NFC is in theory responsible for encouraging agricultural production with a system of grain price supports. The price supports are meant to increase production in the terai and thereby make more food available for transport to the hills. This program has had no impact, because the amount of grain the NFC buys is too small to influence the market, and the support price has almost always been lower than the market price. Moreover, prices on the Indian market dominate the terai markets and limit the potential impact of government price interventions.

6.16   Fertilizer Subsidies. The fertilizer subsidy consists of a transport element and a price element - which is to some extent a function of the need to maintain parity with Indian prices. The net effect is that farmers in the terai and Kathmandu probably pay something close to the real cost of fertilizer, while hill and mountain farmers have been paying below cost. As mentioned in Chapter 4, the subsidy has very little impact on the poor, because:

- they do not use fertilizer (because they are not involved in the cash economy, and because they practice rainfed agriculture with limited scope for fertilizer use);

- supplies through AIC are unreliable, and unavailable at key times; and,

- subsidized fertilizer is deflected to India, to Kathmandu, or goes to the better off.

6.17   Under the second Structural Adjustment Credit HMG is starting to deregulate fertilizer distribution, and reducing subsidies to discourage deflection to India, although the total cost of the subsidy has not yet come down. At any rate it is unlikely that _any_ fertilizer subsidy would have a substantial direct impact on the poor - given the constraints on their effective use of fertilizer.

6.18   Credit Subsidies. Through the Agricultural Development Bank of Nepal (ADBN) HMG provides interest and capital subsidies to encourage productive investments in agriculture (mostly for irrigation, cash crops, and livestock). These subsidies have little direct impact on the poor; less than 10% of such credit is estimated to go to small farmers; they generally have neither the access nor capacity to use institutional credit; and these are not the type or scale of investments which the poor undertake.

6.19   **Irrigation Operations**. The Government provides an indirect subsidy by bearing the operating costs of public irrigation schemes. This implicit subsidy probably benefits the poor to some extent, although the poor tend not to be on irrigated land, so the incidence of the subsidy is mostly on the non-poor, and probably on the wealthy (since it varies in direct proportion to area of landholdings). Our calculations show that if they are on irrigated land, the poor could afford to pay irrigation charges, and in the long run the failure to recover costs will hurt everyone, including the poor, as operations and maintenance deteriorate. HMG's policy is to move towards full cost recovery, either through higher user charges or transfer of responsibility for O&M to beneficiaries.

6.20   Grants are also provided for the capital costs of small irrigation schemes. It has not been possible to estimate the total costs of this subsidy, as it is distributed among a number of projects in the budget, but we suspect it is large. While promoting small scale irrigation is a legitimate poverty-alleviation strategy in Nepal (in fact probably one of the most effective), given the very high private returns it is not clear that a capital subsidy should be necessary except for very small farmers. It is recommended that consideration be given to replacing the capital subsidy over time with a credit program, with an exemption for very small landowners, while retaining the technical assistance elements of such programs.

6.21   **Conclusions**. Existing subsidies and transfers do not result in any substantial transfer of benefits to the poor. There may be other reasons for continuing these programs (eg. to accelerate sectoral growth) but they do not appear to be justified on poverty alleviation grounds. If the objective, in the minds of policy-makers, is to alleviate poverty, then these funds would be better spent on other programs. Even the non-poverty objectives may be better achieved otherwise - eg. by increasing civil servants salaries rather than providing subsidized grain, or by credit rather than capital subsidies for irrigation works.

6.22   Some elements of the transfers and subsidies are poverty-related (eg., NFC's general food distribution for remote areas, SFDP subsidized credit) and these could benefit from being strengthened. Wholesale transfer programs are probably not viable in Nepal in the foreseeable future, given the magnitude of the problem and the limited financial resources available to HMG. (For example, with 7 to 10 million poor, and no easily identifiable sub-group of ultra-poor, the cost of a transfer program which supplemented incomes of the poor by only 20% would be about Rs. 3.9 billion annually, or 25% of public expenditure.)

6.23   The transfer interventions which do make sense in Nepal are those which improve food security for the poor (especially in remote areas), and selected transfers to poor groups or individuals for income-generating activities, where these can be effectively identified and targetted - eg., under some NGO and targetted credit programs. Recommendations to strengthen existing programs (NFC, SFDP, PCRW) are made elsewhere. In addition, HMG may wish to consider additional programs for targetted feeding, nutrition supplementation, and public works employment.

6.24   Subsidy programs (especially for foodstuffs) have had a substantial impact on the welfare of the poor in many countries (eg., Sri Lanka, Mexico,

India), although they have operated largely to the benefit of the urban poor. With the emergence of a large number of urban poor in Nepal over the next 20 years the pressure (and need) for such a program will mount. At that point it will be necessary to re-examine the need for price subsidies, but for the moment there is no justification for subsidizing foodstuffs to urban consumers. The remote areas food distribution element of NFC's subsidy is legitimate, and should be refined (see para. 6.45) - possibly utilizing some of the resources which currently benefit Kathmandu.

6.25   Finally, a whole constellation of subsidies, incentives, and payments in Nepal are designed to compensate for the high costs of access in the hills. In our view it would, in general, be more cost-effective to spend these resources on improving physical access (at least in those areas where it is economically viable) rather than on an inefficient conglomeration of subsidies which are difficult to target, costly to administer, and prone to mis-use.

Integrated Rural Development Projects

6.26   There are at least nine on-going IRDPs, at a total cost of over $250 million. Like IRDPs elsewhere, they take an area focus, include a wide range of sub-projects (agriculture, roads, health posts, schools, water supplies, income-generating projects, etc.), and attempt - at least in theory - to integrate them in a regional development strategy. In fact the sub-components are often indistinguishable from general sectoral programs, except sometimes for implementation arrangements.

6.27   IRDPs are not poverty alleviation projects. However to the extent that they are in relatively poor areas of an absolutely poor country, many of the intended beneficiaries are poor. They tend, by design, to focus on smallholders and those with some productive assets - they do not address the policy issues (eg. land reform, employment, tenant's rights, migration) which most effect the landless and rural labourers. In Nepal, instead, they have increasingly focused on "process" issues of beneficiary participation and organization.

6.28   Internationally, the experience with IRDPs has been less than satisfactory - so much so as to call the validity of the approach into question.[2] The main problems have been: the lack of sustainability of project benefits when activities are financed at a higher level than in the economy as a whole; (ii) unmanageable complexity; (iii) excessive reliance on special project implementation groups; and, (iv) failure of the expected benefits of integration to materialize. The IRDP's in Nepal manifest some or all of these of these problems.

6.29   None of the IRDPs reviewed under this study can claim to have been very successful either in promoting rural development in general or in meeting the needs of the rural poor in particular. In all six cases there is an absence of documentary evidence of project effects that goes much beyond the achievement of targets or accounting for inputs. In most cases the projects are able to claim that they achieved their targets within

---

[2]   See for example, <u>World Bank Experience with Rural Development Projects 1965-1985</u>. IBRD, 1987.

budget, but unable to claim improvements in household income or nutritional status.

6.30   Comparing the performance of agriculture in six IRDP areas with the regions in which they are located reveals no clear trends. The best one can conclude is that the performance of IRDP's may have been no worse than that of other development initiatives in agriculture or forestry, but the performance in these sectors during the last two decades has been marked by almost systematic deterioration, which intensive inputs under IRDP's has not arrested. To the extent that benefits were generally postulated in terms of increased agricultural productivity, one might have hoped for more from expenditures of $25 million p.a..

6.31   There have been some particularly successful sub-components (eg. the Pakribas Agricultural Centre under the Kosi Hills project, financed by ODA); and the design of more recent IRDPs reflects some learning from experience - both with more appropriate implementation arrangements, and many fewer components.3/ However, in the absence of any clear indication of success, and given the very high levels of expenditure, one needs to question whether or not it is more effective to focus instead on: (i) strengthening overall sectoral programs (eg. agriculture, forestry, health); (ii) financing essential infrastructure; (iii) building up the capacity of line Ministries to deliver the services for which they are responsible (including building in the useful "process" reforms being sponsored under some IRDPs); and, (iv) undertaking the necessary countrywide programs and policy reforms which will affect the poor (eg. population measures, developing an appropriate technical package for hills agriculture, investing in road access and education).

6.32   It is recommended that HMG consider adopting this approach rather than pursuing new IRDPs, or, if it is felt to be desirable to focus efforts on some less-developed areas because the strengthening of general programs will take too long, then HMG should consider limiting IRDPs to a restricted number of activities (probably irrigation, and road and trail building - factors which can make a quantum difference to productive potential).

Other Intensive Agriculture Programs

6.33   The Pocket and Block Production programs of the Ministry of Agriculture provide intensive agricultural support services to selected rural areas which have irrigation and which are accessible. These programs are intended to accelerate general agricultural growth, rather than to help the poor specifically. In the areas covered, Ministry staff are supposed to construct cropping plans in collaboration with local farmers and ensure the timely availability of inputs. Local studies commissioned for this report show that in areas where adequate staff is available, substantial increases in production can occur. However, staffing levels in many areas have not been adequate and the Ministry has not been able to insure availability of inputs. It is not clear that these programs are more effective than a well functioning general extension program, and they have concentrated on areas

---

3/ Dhading (1985), for example, relies on local implementation capacity, and Seti and Mechi (1987) both focus only on irrigation and roads and trails.

with superior resource endowments, drawing resources away from poorer areas. Given the uneven experience in implementation it may be better if these programs were discontinued and funding and staff shifted to improving the general extension program.

6.34   The Lumle and Pakhribas Agricultural Centers, both funded by ODA, have developed innovative methods for integrating research and extension services so that the focus of research is relevant to local farmers. Both Centers stress on-farm varietal testing and frequent contact between researchers and farmers. The high levels of staffing needed may limit the ability of these Centers to expand their areas of coverage. In addition, the areas currently covered are somewhat better endowed in terms of agronomic characteristics than many hill areas, and this may have contributed to the success of the Centers. However, the emphasis on frequent contact between farmers and researchers is the most effective method of developing agricultural technologies relevant to local conditions.

Food and Feeding Programs

6.35   The Nepal Food Corporation attempts to reduce inter-regional food imbalances by buying food in surplus areas and selling it at subsidized prices to the poor and government officials in food deficit areas. Political considerations channel most NFC-distributed food to the Kathmandu Valley, despite the relative lack of poverty in the area and the well established market system. In remote areas, NFC food is flown in at great expense, and even in less remote areas the transport costs are heavily subsidized.

6.36   The food distribution program has not had a significant impact on the food security of the poor. The quantities distributed have been too small, usually less than 15% of the deficit in areas outside the Kathmandu Valley, many of the beneficiaries are government officials, and large amounts of the grain goes to people too wealthy to meet eligibility requirements. Despite the problems with this program, the need for it is increasing as population growth continues to outstrip agricultural production. The private market will not provide sufficient food in remote areas in the near future because effective demand is too low at the prices private traders must charge to cover transportation costs.

6.37   The NFC is starting a local grain storage program to try to reduce seasonal price fluctuations by giving poor farmers the opportunity to hold grain after harvest until prices rise. The program centers on organizing poor farmers into savings groups with rotating loan funds, so that they can escape the necessity of selling their crops immediately after the harvest in order to repay debts.

6.38   Food-for-work projects supported by the World Food Program appear to be more successful in reaching the poor than the NFC food distribution program. The potential impact of this program on food security is quite large, although at the moment the total food distributed as wages is less than 2,000 mt p.a., compared to the NFC which distributes 34,000 mt/year, and relative to a total average annual deficit of 300,000 tons.

6.39   The Nutritious Food Programme, also supported by WFP, reaches a larger number of people (190,000 in 1987), and distributes a larger amount of food (5,500 mt in 1987), but some aspects of the program have targetting problems and the full development potential of the food distribution program is not utilized. It includes distribution of food to malnourished children and pregnant and lactating women, and institutional feeding programs at primary schools, orphanages, and child care centers.

6.40   The distribution of food to women and malnourished children is the largest segment of the program. Children are screened once a year, and if they are malnourished, they become eligible for supplementary food. Pregnant and lactating women are automatically eligible for food supplements. Targetting is a problem, with the food often distributed irrespective of income, gender, or age, and case studies have shown massive amounts of leakage. If the targetting problems could be overcome, and if the food distribution was accompanied by health and nutrition education, the impact of the program could potentially be substantial.

6.41   The Joint Nutrition Support Programme, under the guidance of UNICEF, has attempted to coordinate the activities of four government ministries on a wide range of activities (nutrition, agriculture, food storage, sanitation, education, women's income generation, etc.) on the grounds that the multiple causes of malnutrition need to be addressed as a group. The program has largely failed because it is excessively complex, because it is not the sole responsibility of any single ministry, and is not a high priority in any ministry.

6.42   The combined impact of all the programs discussed in this section is relatively small compared to the magnitude of the food security problem in Nepal. All the food handled by the NFC and the World Food Program represents less than 10% of the national food deficit, and much of it goes to households that are not food insecure. In addition, most of the programs have not changed long-term food security at the household level. However, if expanded and improved, they could be important parts of household coping mechanisms, either by supplying subsidized food when supplies are low or by providing seasonal employment.

6.43   The main elements of a strategy to improve food security in Nepal need to revolve around improving agricultural productivity and physical accessibility and raising incomes. However, given the length of time this will take and the remoteness of many communities, some directly food-related measures also form a legitimate part of the solution.

6.44   There is some role for distribution of food to deficit areas; however, one has to question both the sustainability of such transfers, and the capacity of the existing NFC framework to deliver food efficiently without leakage. The preferred solution would be to develop the capacity of private traders to supply food as purchasing power increases in hill areas. However this will not happen quickly enough to relieve food shortages for many, and therefore it is legitimate to distribute food, the question is how best to do it.

6.45   The NFC would have to distribute more food than it currently does to have a significant impact. A restructured program would be beneficial if: (i) the supply of food to public servants were divorced from other feeding functions (as long as it remains it will divert effort, and scarce supplies will always tend to be allocated first to public servants); (ii) better geographical targetting is necessary, as a first step the Kathmandu valley should be eliminated entirely from the program; and, (iii) the scope for more effective beneficiary targetting should be investigated - although we remain skeptical of the capacity to effectively target general food distribution in Nepal. (Food-for-work, for example, may be a more effective mechanism).

6.46   Storage schemes to reduce seasonal variations may have some impact. They would have to be accompanied by a credit component since most poor households must sell crops immediately in order to pay debts. However, the experience worldwide with such programs has been discouraging, and the relevance of storage programs is questionable in the absence of increases in incomes. While there may be an element of oligopolistic pricing which could be reduced by more cooperative storage, one must assume that those currently purchasing and storing grain would be willing to sell it within the area, if those living there had sufficient purchasing power.

6.47   Targetted group feeding programs have proved very difficult in Nepal. Most have failed, and Freedom from Hunger, one of the largest NGO's, is shutting down its feeding operations because of the difficulty in targetting. The Vulnerable Group Feeding program has recently been restructured to aim more specifically at the malnourished - it is recommended that the experience with this program be reviewed in about a year's time, along with that of similar programs in other countries before proceeding with further targetted feeding initiatives. It should be noted, however, that all of the staff with field experience in Nepal interviewed by this mission questioned the capacity to deliver any targetted program.

6.48   There is a lot of mileage to be gained by changing feeding practices. A hygiene/nutrition education program, if it could be well designed and adequately staffed, could thus potentially play a useful role. While changing behavior is always a problematic proposition, experience to date suggests people are receptive, and that the messages to be transmitted are not that complex. (They involve not withholding food from diseased persons and women during pregnancy, better use of weaning foods, boiling water, etc.).

Targetted Credit

6.49   Targetted credit (perhaps due partly to the the relatively poor performance of most IRDPs and sectoral programs) has become an important element in Nepal's poverty alleviation strategy. Currently, considerable attention is being focussed on three programs: the Small Farmers Development Program (SFDP), Production Credit for Rural Women (PCRW), and the Intensive Banking Program (IBP). All three combine elements of community development, income-generation activities, and the provision of credit.

Table 6.3: Targetted Credit Programs

	SFDP	PCRW	INTENSIVE BANKING
Interest Rate	15%	16%	16%
No. of Beneficiaries (Cumulative)	15,000 p.a. (78,500)	2,700 p.a. (6,650)	n.a. (62,150)
Disbursements (Cumulative)	Rs. 150 m. p.a. (Rs. 644 m)	Rs. 8 m. p.a. (Rs. 14 m)	Rs. 140 m. p.a. (Rs. 814 m)
Recovery Rate	48%	91%	57%
Average Loan Size	Rs. 1,900 /b	Rs. 2,105	Rs. 13,100
Estimated Delivery Cost per Beneficiary \a	Rs. 2,050	Rs. 4,075	Rs. 1,010

a/ Administrative costs only, much higher allowing for provision for bad debts.
b/ May be substantially higher, average lending per beneficiary is Rs. 8,200.

6.50   SFDP, under the Agricultural Development Bank of Nepal (ADBN), relies on a network of facilitators who organize small farmers into groups for credit and developmental activities. It appears that most credit is used for working capital and livestock, and that about 20-30% finances longer-term improvements (eg. irrigation and horticulture). Program design incorporates most of the successful elements of similar programs elsewhere (group liability and savings, intensive technical support), however these appear to have lost impetus in the process of implementation - especially following recent expansion.

6.51   PCRW is a government program under which Women's Development Officers form groups of village women for developmental activities and small-scale business undertakings, for which they then receive slightly subsidized commercial bank financing under the IBP (below). PCRW is not primarily a credit program - it has lent to only 6,500 persons (the vast majority for livestock), and 60% of its costs are non-credit related. It is a general development program for women, which uses credit as an entry point to organize groups as vehicles for such activities as literacy, health and family planning, and small public works. Unit costs are high, although this may reflect naturally expensive mobilization efforts.

6.52   In neither program are interest rates much below average commercial lending levels. Delivery costs are very high, and (with the exception of PCRW) recovery rates, while no worse than for many other types of lending in Nepal, are not high enough for the programs to be financially self-sustaining. HMG currently supports both programs by utilizing concessional financing from donors (mostly the ADB, IFAD and UNICEF).

6.53   Despite the weakness of evaluation reports in distinguishing between credit-related benefits and general income increases, it appears both programs are having some impact on incomes among the moderately poor. Case studies of borrowers however reveal many instances of business ventures which fail, or use of funds for non-productive purposes. Moreover, experience from other countries (as well as information on landholding sizes among SFDP clients) indicates that credit-based approaches work best for the "middle poor" - those within say, two deciles of the poverty line, who have some assets, skills, and their socio-economic connections sufficiently intact to take advantage of income-earning opportunities.

6.54   The two programs currently reach an estimated 5% of the population below the poverty line. There are serious questions about the extent to which they can expand effectively to reach a more significant proportion, because their success is dependent on the intensive use of dedicated field staff, who are in short supply. Already there are signs that both programs are suffering from too-rapid expansion, and inadequate training of new organizers. It is recommended that: (i) both programs should consolidate before further expansion; and (ii) training of group organizers should be strengthened.

6.55   The Intensive Banking Program is different from the other two in that it involves HMG directing the commercial banks to lend to particular groups and purposes. Banks are required to lend 25% of total advances4/ to the productive sectors (broadly defined), and 8% to "Priority Sectors" (eg. small agriculture and cottage industry), of which 60% is to be directed to families below the poverty line. Banks are also expected to provide some of the development functions provided under the other two programs. While there has been some success in raising incomes of the moderately poor, most credit has gone to larger cottage industry projects. The program is characterized by non-compliance by the banks, because the unit costs of small loans are sufficiently high that the banks are losing money on them, banks are not reimbursed for the costs of non-banking services, and the penalty for non-compliance (holding of deposits in non-interest bearing accounts with the central bank) is not a sufficient deterrent to ensure compliance.

6.56   Banks have traditionally been extremely conservative in Nepal - lending only against very safe investments (goods in warehouses, land, jewelry). It is legitimate for HMG to try to encourage the financial sector to undertake more productive lending, and to hasten its spread into rural areas. However, experience worldwide raises doubts as to the sustainability of directed credit; because: (i) it is inefficient, if one is trying to develop a commercial banking system, to then interfere with commercial decision-making; (ii) if banks are not interested they will ultimately not do it; and (iii) if credit is subsidized, it tends to be captured by the non-poor. In Nepal these problems are compounded by the extreme weakness of the banking system, which makes it more difficult for banks to bear the costs of delivering non-credit services required under IBP (and administratively less able to do so).

4/ Defined as investments plus loans.

6.57   There has been a tendency throughout to combine legitimate (but largely non-credit) objectives of group formation, organization of the poor, and assistance in project preparation, with the provision of credit. The use of small-scale credit-led programs as an entry point to reach the moderately poor is legitimate, but for larger scale operations, such as those under IBP, a more useful approach may be to provide these services under the auspices of some developmental program, which would, among other things, help its clients secure credit (either from SFDP, commercial banks, or non-formal sources). It is recommended that for commercial scale investments, HMG divorce the purely commercial banking functions from these developmental functions, and that commercial banks not be expected to deliver developmental services. (At a minimum, if IBP is to continue, HMG must explicitly subsidize the non-credit costs of the program). For micro-level investments which affect the poor, it is recommended that expansion be concentrated on SFDP and PCRW, and not involve the commercial banks. One option might be establishment of a "Poverty Bank", or development fund which could incorporate financing for both programs outside of the commercial banking system.

6.58   Under a recent variant of IBP (the Lead Bank Scheme), one bank in each area is also to help coordinate the delivery of other developmental services. A major expansion of the scheme is proposed as part of the Basic Needs initiative - which would involve the banks increasing by ten-fold their lending to the priority sectors, and directing a quarter of their lending to the poor. In early 1990, HMG announced a further new initiative, whereby injections of credit to poor families (mostly through the commercial banking system) would be combined with local and national-level coordination of supporting development services in a major poverty alleviation effort. This proposal appears to be a combination of the SFDP and Lead Bank Scheme approaches.

6.59   It is questionable whether the strong emphasis on credit as an income-raising strategy is warranted. It is predicated on the assumption that credit (rather than the lack of investment opportunities) is the binding constraint; and that to the extent that such opportunities exist, the poor will be in a position to take advantage of them (which they seldom are). There are severe limitations on productive investment opportunities (lack of transport, markets, and of a monetized rural economy) - any one of which would be more binding than the investment-financing constraint. Certainly credit is one of a number of important inputs, and HMG should continue to ensure it is available as part of its small business development strategy, however to the extent that such limited opportunities exist, it is unlikely to be the poor who will be in a position to take advantage of them.

6.60   Furthermore, we have seen (in Chapter 3) that the problem of indebtedness among the poor in Nepal is one of insufficient incomes to meet consumption expenditure - in these circumstances the insufficiency of credit is a symptom of poverty, not its cause. This is confirmed by disturbing stories suggesting that many SFDP and other loans intended to raise incomes are used to finance consumption or repay previous debts, rather than to finance productive investments. There are three ways in which credit may be of use to the poor: some micro-investments (eg. purchase of a goat or buffalo) can improve their incomes marginally; providing formal credit may reduce the consumption debt-service burden, and reduce dependence on landlords or employers; and small amounts of credit for working capital may be useful to small farmers (who could not otherwise get access to it on affordable terms). These are arguments for providing small amounts of highly directed credit which will have a useful but limited welfare effect. Such micro-credit should not be confused with more substantial poverty alleviation through the massive provision of credit, such as is currently being proposed.

Employment Creation Projects

6.61   There are no large-scale employment programs as such in Nepal; the Special Public Works Program (SPWP, sponsored by ILO) is focussed on small irrigation projects with a labour-intensive emphasis, and the Food-for-Work program is part of the WFP's food distribution efforts. SPWP has employed perhaps 10-20,000 rural people in construction, many of them poor - but the main benefit has been from irrigation improvements, which tend to benefit households in direct proportion to their landholdings. It has had success in raising the labour intensity of rural works, achieving rates of 56% of expenditure on wages, as opposed to 30-40% otherwise - but at the cost of intensive supervision and support for very small local petty contractors. The project has also experienced equity problems in the use of "voluntary" labour, because despite being designed to maximize beneficiary participation, in practice the workers involved in construction are often not from the same groups as those who benefit from the irrigation works.

6.62   SPWP is useful as a tool for increased irrigation and agricultural intensity - but in this respect not differentiable from intensifying irrigation in general. As a direct employment program it is unlikely to have a major impact on poverty at its current level, as it creates only an average of 1,350 full-time construction jobs. Wider use of the same approach - through an expanded SPWP or otherwise - could however have a significant impact on construction employment.

6.63   Irrigation (and other rural works) are probably more likely to benefit the not-so-poor - careful attention needs to be paid to selection of sub-projects, and consideration given to compensating measures which spread the benefits more equitably (the example of a water tax, combined with paid maintenance workers, is a good one).

6.64   Subject to a review of cost-effectiveness, the program probably should be expanded as rapidly as its absorptive capacity allows. The approach should be replicated in other areas - most notably on rural roads, with attempts to refine the system of small-contractor use. Water taxes or user charges should be introduced once schemes are completed, and the proceeds used to pay wages to poorer villagers for operations and maintenance.

6.65   The Food-for-Work program (see para. 6.38) provides about 1 million days of employment per year (6,600 full-time equivalent jobs). It pays an in-kind wage, the value of which is primarily attractive to the poor. In addition the work, building roads and trails, is offered during the dry season, when other employment opportunities are scarce.

6.66   Some useful lessons emerge from rural employment schemes elsewhere. In particular, the experience with programs in India suggest that they must provide employment on a significant scale, and on a reliable basis, to have a meaningful impact on poverty status. The Employment Guarantee Scheme in Maharastra has been particularly successful by providing an employment guarantee which has broadened the material options for the poor, and reduced the extreme vulnerability which is a hallmark of poverty.

6.67   It incorporates a number of instructive features. First, the wage has been intentionally kept below prevailing market rates - dissuading healthy adult males from participating except during seasonal lulls or when disasters have hit. Labourers, who are invariably drawn from landless and marginal cultivator households, are primarily (at least 60%) women who are attracted by the proximity of work sites, by the self-pacing nature of the work (piece-rates apply), by the pattern of working in groups, and by the practice of paying wages directly to laborers without intermediation by contractors or the panchayat. It also involves extensive monitoring and supervision of work sites.

6.68   However, what distinguishes EGS from other rural employment programs is the statutory work guarantee which represents precisely the kind of insurance which is needed by the poorest. The susceptibility of very poor households to sharp reductions in income and loss of assets can only be counterbalanced by a non-alterable fallback arrangement of some type.

6.69   The feasibility of such a broad-based program in Nepal is questionable, both because of the high costs and because of limited administrative capacity. However, some version of the EGS may be useful in particularly poor areas, and elements of the approach could certainly guide design of a labour-intensive rural works program. It is recommended, that following preparatory analytical work (see para. 4.44) HMG undertake a program of public works employment aimed particularly at the poor which incorporates features of the SPWP and food-for-work, and EGS programs, as well as drawing on experience of similar programs elsewhere.

Income Generating Projects

6.70   There are many small income-generating projects focussed on activities such as tailoring, bee-keeping, sericulture, and handicrafts. They are usually wrapped up with more general community development efforts of NGOs which combine agriculture with some health care, education, and small business projects within a few villages or districts.

6.71   In many cases they involve very intensive support (eg. helping to supply inputs, providing the technology, and marketing of outputs). Reliance on such extreme levels of external support raises questions about the sustainability of these projects. While such intensive interventions may dramatically affect the incomes of individual poor households, they do not address the more general issue of constraints on the overall expansion of non-formal sector activity (see Chapter 4).

6.72   The aggregate numbers being reached are small (although not insignificant). There are almost no quantitative evaluations of these programs. There are some reports of localized successes - often where access has recently been improved - and many accounts of financial failures. It may be indicative that the income-generating components of several IRDPs have been dropped as non-performing. The areas of success appear to be those associated with forming the poor into groups, promoting self-reliance, and the like.

6.73   The intensive use of resources to reach small numbers of families suggests that this is not a major element of poverty alleviation strategy. However it is possible that the high costs of supporting these enterprises represent transition costs associated with the change to a cash economy - which will decline as people become more accustomed to starting small businesses, and as input and marketing systems become more developed. These are the sort of interventions in which NGOs are particularly useful at targetting, although their capacity to pick financially sound projects appears weak. It is recommended that consideration be given to establishing a fund for financing small income generating projects for NGO implementation with strict beneficiary criteria, and a small technical group to help appraise the financial feasibility of sub-projects.

6.74   A wide range of programs also provide training for self-employment or to provide marketable skills. The largest are Labour Training Centers run by the Labour Department and the Training for Rural Gainful Activities program (TRUGA). It is not obvious that they are particularly selective of the poor, rather than of the population in general. Most evaluations report only limited success - mostly because of the lack of employment opportunities for those trained. TRUGA, for example, while having success in integrating into local villages, has trained only 1,000 people over four years. A multitude of small programs provide training specifically for women - in the usual areas of sewing and handicrafts - with little analysis of whether there are opportunities for them to profitably use these skills, and equally little employment of those trained. While skills training potentially has a contribution to make to long-term growth in Nepal, the projects tried to date would need better design and implementation to have an impact on the incomes of the poor.

C. Institutional Issues

Civil Service and Service Delivery Issues

6.75   The civil service is still in the early stages of development - prior to the 1960's its functions were limited to tax collection and maintaining public order. A wide range of problems plague the delivery of public services. They include:

- Inadequate Renumeration and Incentives and Poor Supervision - which result in high absenteeism, low morale, and under-staffing in rural areas;

- Promotions and Postings Practices - based on narrowly - defined performance criteria, and influenced by patronage connections;

- <u>Inadequate Operating Funds</u> - coupled with unwieldy funds release mechanisms - which result in inadequate supplies and travel - especially to outstations;

- <u>Excessive Centralization</u> of resources, staff, and decision-making;

- <u>Attitudinal and Communications Problems</u> - which prevent effective contact between public servants and illiterate peasants.

6.76   These affect all HMG activities, they are not specific to poverty-related programs. However, they have a disproportionate impact on: (a) interventions which require complex service delivery, and (b) the quality of services in outlying areas; both of which are particularly important in reaching the poor (services such as health, family planning and agricultural extension are especially hard hit).

6.77   The problems of public service appointments, posting, supervision, etc. - while deeply entrenched - can probably be ameliorated by development of more systematic personnel and administrative systems. <u>It is recommended</u> that HMG undertake a systematic review and reform rather than addressing them sector-by-sector. In the meantime, there is an acute shortage of local service delivery capacity, and this needs to be borne in mind in designing poverty-related interventions.

Decentralization and Popular Participation

6.78   There are two elements of "process reform" which have a potential bearing on poverty alleviation efforts in Nepal. The first is greater self-reliance, participation in development initiatives, and organization among rural people. The second is decentralization of administrative responsibilities to District and local level, along with the establishment of local representative councils (pursuant to the Decentralization Act of 1982).

6.79   With respect to the first, many engaged in development in Nepal believe that the predominantly top-down approach followed to date has resulted in few benefits, selection of inappropriate projects, and an erosion of self-reliance; and that community mobilization can lead to more effective interventions. Some success has been had in forming groups (eg. of forestry and irrigation users, and under some NGO operations and IRDPs) - which allow the poor to more effectively demand and utilize services. While such initiatives can be very useful, their success: (a) is often dependent on the presence of a skilled organizer, and (b) has been limited to the extent that they do not challenge the existing political (and economic) order. They yield useful lessons for the design of other interventions, but as a broad poverty alleviation strategy their effectiveness may be limited to the extent that the binding constraints (i.e. the distribution of real political power) is not being lifted.

6.80   <u>Administrative decentralization</u> can help the poor to the extent that resources are moved closer to poor beneficiaries. So far, however, there has been a natural reluctance to transfer real control over staffing and budgets. In addition, if revenue generation for decentralized functions is transferred at the same time, the poor may lose - as poorer Districts, which have the weakest resource base, will be less able to support services or finance local projects. In addition, placing responsibility for delivery of complex services at the level of government with the fewest resources and the least experience of planning and implementation seems unreasonable.

6.81   The transfer of political power to local levels can potentially increase responsiveness. However, experience with local government elsewhere in the world does not suggest that it is particularly more representative of, or responsive to, the needs of the poor. (Nor does experience in Nepal suggest that this is the case here (see Chapter 3)). Decentralization is a necessary stage in the evolution of administration in Nepal, but it should not be counted on to have major poverty-alleviation effects.

Non-Governmental Organizations

6.82   There are a large number of international NGO's active in Nepal (although mostly not engaged directly in income-generating activities), while the national NGO sector is still in a fledgling state. There have traditionally been a wide range of indigenous self-help groups and community service organizations - but Nepal does not have the range of local developmental NGO's found in neighboring countries.

6.83   The activities of some 150 NGO's were surveyed for this study; most of them are engaged in community services or the provision of health and education; although some (about 15) are increasingly engaged in income-generating activities. Among the most active are Action Aid, Save the Children, CARE, Lutheran World Services, World Neighbors, Redd Barna, the Association of Craft Producers, and Integrated Development Systems, a consultancy-cum-development agency.

6.84   The benefits of NGO's are that they can try more innovative solutions, are not constrained by government staffing and financing procedures, and can work intensively at the local level. This intensity is also a weakness, in that their capacity to operate on a large scale is limited by the availability of dedicated staff. Major donors, on the other hand, have the resources to finance programs large enough to have an impact on poverty, but not the capacity to design or implement the intricate service delivery and social mobilization mechanisms needed for them to succeed. There are thus opportunities for the greater involvement of NGOs in executing large-scale projects, as well as in enhancing the "process" aspects of implementing government's general development programs.

6.85   The principal benefit of NGOs may be that they can operate free of the political and social constraints which limit the capacity of government programs to help the poor. With respect to poverty alleviation efforts, caution should be exercised, because to the extent that existing economic interests are threatened by organizing the poor, it seems unlikely that NGOs will be allowed to operate freely.

6.86   NGO activities are governed by the Social Services National Coordinating Council (SSNCC), through which both funds and approvals for NGO activities are channeled. HMG's attitude towards NGOs has been ambivalent;

the SSNCC provides a central monitoring and coordination function, but is also an agent of control. This reflects a legitimate concern to prevent a plethora of agencies (many of them foreign) operating in the countryside unpoliced, but it may also limit the effectiveness and flexibility of NGOs. There is room for revamping the framework for NGO coordination, especially if NGOs are to play the active role prescribed for them under the Basic Needs Program. It is recommended that a review be undertaken of (a) the regulatory framework for NGOs, taking account of arrangements in other countries; and (b) the scope for establishing a fund through which modest levels of donor and NGO financing can be channeled to local NGOs; and assistance can be provided in strengthening their administrative and project selection capacity.

VII. COUNTRY STRATEGY IMPLICATIONS

7.1 This chapter consolidates the findings on the potential contributions of the various sectors to raising incomes and absorbing labour. The objective is to point to areas in which to concentrate efforts, and to frame the bounds within which HMG needs to design a poverty alleviation strategy. Specific policy and program recommendations are discussed in the following chapter.

7.2 Nepal is a very poor country, with a relatively even distribution of income and land, therefore there is limited scope for poverty alleviation through redistribution or welfare measures, the need is for targetted growth. However at current rates of population growth (2.7% p.a.), and any expected levels of economic growth, there is no hope of seriously reducing the number of poor. Curbing population growth is thus one of the central elements of poverty alleviation strategy for Nepal. To reduce the incidence of poverty to say 20% by 2010 (even with optimistic assumptions regarding GDP growth) would require reducing the population growth rate to around 1.7% p.a.. In the absence of this sort of achievement, any other income increasing measures will be meaningless.

Agriculture

7.3 We have seen that about two-thirds of rural hill households and just under half of terai households could never be expected to rise out of poverty on the basis of agriculture alone - their holdings are too small to do so. It is incorrect to keep thinking of these people as poor farmers, rather they are poor households, who happen to own some land. Their coping strategies thus have to depend on a range of interventions. This is not to say that agriculture is not important, especially in the terai. The poor receive half or more of their current income from agriculture, so even the increases of 20-30% in output which are achievable would have a substantial impact. In addition, they are disproportionately affected by the demand for agricultural labour. Equally importantly, experience in other countries shows that while agriculture itself may not be the solution to rural poverty, reasonably equitable growth in agricultural incomes leads to very significant expansion of non-farm rural employment.

7.4 Furthermore, it should be remembered that in a heavily agricultural economy which is not heavily taxed, and where the agricultural transition has scarcely begun, the social safety net needs to be built of resources found in farming itself; resources are not yet being transferred out of the sector for welfare purposes, that is, to take care of people crowded out of the economy for lack of adequate resources, or because of age or infirmity of one kind or another. So while these small plots may not produce much, their major use is for the succor of the families that live on them in a job-scarce economy. In economies such as that of Nepal agriculture must serve not only the functions we usually attribute to farming, but also as a social safety net.

7.5 About a third of the agricultural poor in the hills have enough land to potentially raise themselves above the poverty line through agricultural improvements (but only if their land is irrigable). In the terai, perhaps 40% of the poor have large enough holdings (say about 0.6 ha.) to produce above-poverty household incomes, if fully irrigated. Perhaps another third of the agricultural poor would, with foreseeable technical improvements, produce a significant share of poverty-line incomes from agriculture alone. Labour absorption as the result of transformations in agricultural technology could potentially result in employment for about an additional 250,000 persons in the hills and one million in the terai.1/ It can be assumed that most of these jobs will absorb the poor (or family members of non-poor households who would otherwise become poor as a result of land sub-division). In sum, the best that could be hoped for from agriculture in the foreseeable future might be to support another four or five million of the poor (about 700,000 households). Scope for land redistribution is limited, especially in the hills; however a redistributive reform in the terai - if carefully targetted - might raise 400,000 households out of poverty.

Off-Farm Incomes

7.6   We have seen (Chapter 4) that the formal sector might be expected to generate an additional 1.2 million jobs by 2010, although outside construction, few of them absorb the poor. (Bearing in mind that at current wage levels and family sizes, off-farm jobs do not generate sufficient income to raise the poor out of poverty unless several family members are employed.) Hydro electricity exports have the potential to increase GDP by about 10%. However beneficial income effects depend on following consciously redistributive policies, and taking very specific steps to counter terms-of-trade effects which would be biased against the poor. The scope for macro-economic adjustments which would significantly impact on incomes of the poor appears to be extremely limited. The only caveat is that if India were to undergo a major liberalization of its trade regime, then Nepal, by following suit, could potentially open up much faster formal sector growth.

7.7   The informal sector can at best follow growth led by the other sectors - with allowance for a catch-up effect due to the current low level of monetization. Measures tried in other countries to "unleash" self-employment in the non-formal sector (eg. credit, skills training) may not be very effective due to difficulties of access and the lack of an effective cash economy - although the scope in urban areas and the terai is growing. Analysis in Annex II.5 suggests that about an additional 1.8 million persons might be absorbed in informal activities within the next twenty years.

Aggregate Potential

7.8   Table 7.1 shows that if our (relatively optimistic) expectations are met regarding growth of GDP and formal sector employment, and intensification of agriculture, then labour absorption can just about keep

---

1/ See Annex II.6.

pace with labour force growth. The absolute number of under/unemployed (as a proxy for the number of poor) would increase by about one million, although the proportion of the LF underemployed is substantially lower than today. If, on the other hand, GDP growth is slower, and agriculture performs only slightly better than it has done to date, then the number of unemployed will increase dramatically - perhaps more than doubling. Finally, the last line in Table 7.1 corresponds to an absolute transformation of the Nepalese economy, such as has been experienced in Malaysia or the Republic of Korea. Such a transformation, however, appears unlikely at this stage.

Table 7.1: Possible Labour Absorption - 2010
(millions of full-time equivalent jobs)

	Agriculture/a	Non-Formal Sector	Formal Sector	Total	LF	(Under/ Unemployed)
1990	2.5	1.2	0.8	4.5	7.9	(3.4)
Lower Bound	3.1	2.1	1.4	6.6	13.6	(7.0)
2010 Medium	4.2	3.0	2.1	9.3	13.6	(4.3)
Upper Bound	4.9	4.3	2.9	12.1	13.6	(1.5)

a/ Full-time equivalent jobs in cropping, at 180 days/year - actual number of persons engaged is much higher due to shared underemployment, and effects of non-crop labour requirements.

7.9   Table 7.2 illustrates, in very rough terms, the impact various developments could have on reducing the number of poor. What is notable is that: (i) no one set of measures stand out; (ii) even achieving these results is dependent on successful new initiatives (eg. in expanding irrigation, intensifying cropping, etc.); (iii) most of these measures will only have the indicated effect if intensive efforts are made to ensure that the poor (as opposed to the non-poor) benefit; and (iv) the potential impact of redistributive measures is not insignificant - however the capacity to effectively target them is questionable.

Table 7.2: Potential Impact of Various Developments on the Incidence of Poverty
(Number of poor households potentially raised above poverty
line by various events)

Increased Agricultural Production On Own Land /a		Increased Agricultural Employment /c		Opening up Terai Lands /d	Formal Sector Employment /e	Informal Sector Employment /f	Terai Land Redistribution /g	Income Redistribution /h
Terai	Hills	Terai	Hills					
0.15 m.	0.25 m.	0.25 m.	0.13 m.	0.4 m.	0.25 m.	0.6 m.	0.4 m.	0.3 m.
(0.23)/b	(0.1)/b							

---

a/ Currently poor households with enough land (0.5-1.0 ha.) to produce above poverty incomes from cropping alone.

b/ Poor households with enough land to produce a substantial share of poverty level income (say 50% +) from cropping alone, with improved technologies (15% of hills poor, 20% of rural terai households).

c/ On holdings larger than 1 ha.; as a result of irrigation and intensified cropping; 2 jobs per household.

d/ Conversion of 400,000 ha. of remaining forests to farm land - if distributed equally in 1 ha. lots among poor households.

e/ Incremental growth to 2010 - assuming 80% of construction jobs and 20% of others absorb the poor; 2 jobs per household.

f/ As with (e); assuming 70% absorb the poor.

g/ Redistribution of holdings above 4 ha. (20% of terai land) - if redistributed in equal plots exclusively among poor households.

h/ 15% tax on incomes of the top 10% (20% in urban areas) if distributed among those in the bottom 5 deciles.

7.10   There is then no easy poverty alleviation strategy for Nepal. The solution must lie in growth. However growth fueled by the modern sector (eg. cash-cropping, urban-based, industrial) has tended not to reach the poor, except in selected cases. Such a growth strategy thus needs to be focussed: on raising lower-level (eg. subsistence) agricultural productivity, on labour-absorbing measures, and on "enabling" measures - setting the preconditions which would allow the poor to help themselves - by equipping them with education and skills and by providing infrastructure. Particular proposals are discussed in the next chapter.

7.11   It must be emphasized that if the Government acts decisively there is scope for a major impact on the incidence of poverty through public policy. For example, an effective program to curb population growth would reduce the number of poor by at least 5 million by the year 2010. Similarly a combination of measures to raise the agricultural productivity of smallholders, and to increase agricultural employment on larger holdings (mostly through irrigation), coupled with selected rural works employment and redistributive measures (see Table 7.2) could in aggregate produce above-poverty line incomes for over 2.5 million poor households, effectively eliminating most absolute poverty by the year 2010.

7.12   Conversely, failure to act on some of these key areas would likely lead to a catastrophic increase in the incidence of poverty. In particular, if population growth is allowed to continue unabated at its current level, this will likely add another 4-5 million persons living at close to starvation levels by 2010, and a further 10-15 million in the following twenty years.

7.13   While effectively curbing population expansion is central to poverty alleviation efforts, there can be little hope of major advances as long as economic growth averages below 3% per annum. Nepal faces formidable natural disadvantages, and in this restricted environment, government policy must be particularly effective to have a meaningful impact. The Government will need to pursue a more active program of reforms than it has generally over the last three decades, and in particular it needs to follow through firmly on the implementation of reforms and on the delivery of programs.

Some Emerging Issues

7.14   The numbers of the poor could easily increase over the next two decades, and the conditions under which they live could simultaneously worsen. Firstly, the rural land base is saturated in the hills - average plot sizes among the poor are already below minimum economic size, and with further intra-family subdivision, many non-poor families will join the ranks of the poor within one generation. In the terai, average holding size is projected to fall by 50% as the result of population growth - perhaps marginalizing 25% of currently non-poor landholders. Secondly, forces are in place for a deterioration in the distribution of income which usually accompanies monetization and urbanization. We have seen (Chapter 3) that if the distribution of income were to shift to one more typical of developing countries, then at expected levels of GDP and population growth, the bottom 60% of the population would have incomes substantially below poverty levels. Finally, Nepal will soon witness the emergence of an urban under-class who lack even the weak asset base and social networks which currently support many of the poor. Only imaginative counter-measures can prevent this decline in conditions of the poor, emphasizing the need for the type of "balanced", labour-absorbing growth referred to above.

7.15   It will take 15 or 20 years for family planning measures to have a major impact on population pressure. In the meantime a large proportion of the poor will continue to cope through migration - to the terai, and increasingly to urban areas. Managing this transition needs to be a key element in country strategy. The rural terai has the potential to support a substantially larger population (about double its current level), but only if: (a) further forest lands are opened up, and (b) steps are taken to raise agricultural productivity, mostly through irrigation. Even then saturation will be reached within 20 years, and thereafter all incremental growth (about 650,000 persons per year) will have to migrate to urban areas or to India. The challenge will be to provide basic urban services, especially those used by the poor, while resisting the pressure to provide benefits for the urban non-poor, and avoiding implicit subsidies which will increase the incentive to migrate. One option would be to focus on middle-sized settlements (eg. roadhead bazaars) which can provide growth loci for rural regions.

7.16   Migration has been an adjustment mechanism in the hills for some time. Incomes there have stabilized at just about survival level. Unfortunately, measures to improve incomes in the hills may do little more than slow the pace of out-migration. More broadly, with the open border, it is likely that real wages will equilibriate at about the level of northern India. Income growth for the Indian poor is thus a major factor in raising incomes of Nepal's poor.

7.17   Twenty one of Nepal's 75 districts have no road, and outside Kathmandu only two districts have minimal road densities of 20 km. per 100 sq. km. In many areas providing improved access is the single most important intervention HMG can make, following which market forces alone can provide tremendous increases in both agricultural opportunities and off-farm employment. However, given the very high costs of road building and maintenance, care has to be taken in defining: (i) a subset of areas in which improved access is economically justified; (ii) those not accessible at reasonable cost in which it may be efficient to provide a minimal package for increasing incomes in-situ (for example, through improved food production); and, (iii) those from which out-migration should be encouraged (see para. 8.09).

7.18   Developing effective education strategies is doubly important for poverty alleviation in Nepal. Given Nepal's limited resource base, developing a skilled and trained workforce will be central to any long-term development strategy. This is admittedly a long-term proposition, which will take several generations, however a start has to be made now. Fortunately, basic education also makes an important near-term contribution to the earning prospects of the poor, as well as having beneficial fertility and hygiene effects.

7.19   It will take time for population programs and productivity improvements to affect incomes on a significant scale. In the meantime there will remain a very large number of absolute poor (perhaps a pool of 5-10 million over the next twenty years), it is legitimate to consider a sustained program of support for them. It is not very meaningful to make the distinction (useful in many countries) between the poor and a sub-class of the "ultra-poor". It is therefore not particularly useful to target the poor for relief measures - there are too many of them. The emphasis instead has to be on balanced growth through effective general programs - with efforts to ensure that they reach the poor. There are some exceptions - particularly related to food security, and these are discussed in the following chapter.

7.20   Projections show that food balances are likely to get worse, at least over the next ten years, with little chance of there being sufficient purchasing power to import the shortfall. Individual food insecurity is ultimately a function of insufficient incomes. In Nepal, however, due to high transport costs and the absence of marketing networks many people in many areas would not be able to purchase food at any expected income levels in the near future. There are sufficiently large numbers of people in these remote, food deficit areas that efforts designed to address food security per se are a legitimate part of any poverty alleviation program.

7.21   There is also scope for improving the welfare of the poor at existing income levels, by: (i) improving the retention of calories consumed but not utilized because of parasites and other infestations; (ii) reducing the incidence of common, preventable diseases; (iii) changing feeding and hygiene practices; and (iv) reducing time spent in gathering water, fuel and fodder.

7.22   Measures which focus on women in particular will probably improve the equity aspects of growth, because women predominate in poor households, because they are disproportionately not participating in the labour force, and because they tend to be most concerned with the family welfare aspects of expenditure. Given the limited labour market opportunities, measures which concentrate on on-farm income earning opportunities for females are probably the most fruitful. Women are also the key to successful population control initiatives, and to the behavioral changes (eg. with respect to hygiene and feeding practices) which can improve living conditions at existing income levels. Female education is important for the contribution it can make in all of these areas.

7.23   Finally, we have seen that the poor could fare better under existing economic conditions by freeing up the political and social environment at the village level to allow more self-reliant activity; and by eliminating some of the more exploitative aspects of labour and construction contracting arrangements. Also the current tenancy system provides disincentives both to maximizing output, to equitably sharing the costs of inputs, and to allowing security of tenure.

VIII. POLICY CONCLUSIONS

A. Overall Strategy

8.1   With its substantial resource limitations and a disadvantageous location, Nepal faces a difficult task in raising incomes. That is not to say that the situation is hopeless - the foregoing analysis demonstrates that growth balanced across the key sectors could be sufficient to raise many of those who are currently poor out of poverty. It is important to recognize, however, that there is no easy solution to poverty in Nepal. In the previous chapter we have seen that there are a constellation of measures which in combination with one another, <u>and if effectively implemented</u>, could in aggregate have a substantial impact in reducing the incidence of poverty.

8.2   The main thrust of a poverty alleviation strategy in Nepal needs to revolve around:

   (i)   a national population campaign;

   (ii)   agricultural intensification - particularly through improved irrigation;

   (iii) increased rural access in the hills and the terai; and,

   (iv)   education - focussing in particular on the lower levels.

Important incremental gains will also come from a supporting program of:

   - Selected off-farm income-generating activities - particularly a program of rural public works, and small amounts of focused credit;

   - Some relief measures - especially to improve access to food; and,

   - Measures which improve the quality of life in the absence of income increases, particularly hygiene and health related ones.

In a number of areas which hold promise, additional analysis is needed before proceeding, particularly on strategies for managing population movements and rural access, for improving food supplies in remote areas, and for agrarian reform. The analysis has also shown that in some areas government interventions will not particularly help the poor - including large-scale use of targetted credit, price controls or subsidies, or industrial sector interventions. Income-generating and skills training projects could have some impact, but would need better design and implementation. The following sections briefly summarize the main policy conclusions by sector; specific recommendations are listed in the Annex 1 which follows.

B. Sectoral Policies

Population

8.3   The need is to implement an effective program of support for temporary methods of contraception - this should be a central nationalpriority. Such a program needs to rely on a large cadre of well-trained and well paid female family planning workers at the village level, to provide continuous support to contraceptive users. Every effort needs to be put into strengthening the institutional and managerial capacity to run such a program; in doing so all necessary financial and managerial resources should be made available without hesitation. At the same time the Government should continue to revitalize and strengthen the established sterilization program.

Agriculture

8.4   HMG needs to pursue two parallel tracks - one, for growth, consisting of traditional agricultural programs focusing on those farmers with enough resources to make use of modern inputs - these will yield the greatest returns, but will bring direct benefits to only about 25% of farmers. Therefore the other track needs to focus, in the hills on marginal improvements in rainfed agriculture and systems largely free of purchased inputs; and in the terai on delivering effective irrigation to small farmers (including a package of tenancy reform, credit, and support for shallow tube-wells). Recent initiatives in community forestry and small farmer irrigation are steps in the right direction, as are changes in the research and extension system which should make it more appropriate to small farmers. These initiatives should be closely monitored - and adjusted if necessary - to ensure that they do in fact reach the poor. There is also a need to deregulate input supply, and encourage development of a network of small agricultural traders.

Employment and Wages

8.5   Absorbing the rapidly growing numbers of landless in employment represents a major part of the poverty challenge in Nepal. Unfortunately there are few obvious measures open to HMG to achieve this. The greatest impact will probably come from measures to increase labour absorption in agriculture - particularly getting irrigation working in the terai. However, the scope for policy measures to increase off-farm employment is limited by the natural constraints to formal sector growth and the open border with India. HMG's main role should be in providing enabling mechanisms: transport and communications infrastructure, general education, and to a lesser extent skills training. A rural public works program will also have some effect. In addition, given the magnitude of the emerging employment problem, at the margin policies should be followed which maximize labour absorption - through following labour-intensive technologies in public works, providing reasonable incentives to labour-employing enterprises, and avoiding incentives which encourage mechanization or capital intensive production.

8.6   The dynamics of the informal sector are not well enough understood to know what measures would help the poor, especially since interventions tried to date seem to have had little effect. Again HMG can probably achieve the greatest impact by providing infrastructure and education. Analytical work should be undertaken to identify whether or not there is a set of more specific measures which could raise informal sector incomes among the poor. On the basis of the preliminary review carried out under this study no obvious candidates emerge.

8.7   There is also little scope for influencing wages, because at the margin they are driven by agricultural productivity, and by the free flow of labour to and from India. The focus instead should be on helping the poor overcome the most exploitative aspects of current labour relations by: selective reform of labour and construction contracting practices, and empowering the poor through group formation, education, and selected credit programs. In the long run the most substantial impact may come from a tightening of the labour market following effective population efforts.

Subsidies, Transfers, and Price Policies

8.8   Existing subsidies and transfers do not reach the poor - they should be re-allocated to programs that do (eg. food-for-work), or where possible re-targetted. Most of the collection of subsidies intended to compensate for remote access in the hills should be abolished and the funds put into improving access instead. There is little justification for price interventions of any kind, most do not benefit the poor, and anyway they are largely irrelevant given the open border - except in the case of foodstuffs for remote areas, and even those are questionable as they often cannot be effectively targetted.

Infrastructure and Spatial Development Issues

8.9   Providing transport infrastructure is one of the most effective interventions HMG can make - however only in some instances. A poverty alleviation strategy for the hills needs to distinguish between areas for which varying degrees of support are justified. The hills should be partitioned conceptually into three categories: (a) areas where road access is economically justified - for which HMG should finance an accelerated program of access; (b) those areas with sufficient potential to justify limited access improvements (eg. suspension bridges and trails, minor roads), and commitment to a continuing program of low level support - perhaps in the form of food distribution, public employment, etc.; in these areas investments in infrastructure should be limited to those with a welfare effect - eg. low-cost water supplies, schools, and health centers; and (c) finally, those areas with no hope of access and where continued expenditure is not justified in the long run - ultimately these areas should depopulate. HMG should undertake only minimal investments in these areas, along with a time-bound program of essential support justified on welfare grounds - but with a timetable for phasing them out.[1]

---

[1] There will be a fourth category, of isolated pockets with high potential which can support a significant population in comfort; solutions need to be tailored for them on a case-by-case basis.

Access to Food

8.10 Improved access to food in remote areas depends mostly on raising incomes and improving physical access, and in some cases outmigration. In many areas this will not happen fast enough to relieve chronic food deficits. In these areas the most cost-effective solution is to produce food <u>in-situ</u>; even this will be a slow process, in the interim limited food distribution is justified. Measures include: (i) encouraging the growth of a private food sales and distribution network; (ii) selected interventions to compensate for remoteness and seasonal variation in food supply; (iii) some feeding of vulnerable groups, if and where they can be cost-effectively targetted; and, (iv) measures to improve nutritional retention at existing consumption levels (eg. through nutritional education and changed feeding practices).

8.11 As a first step HMG and the relevant donors should undertake a food security programming exercise, to include:

- agreement on the appropriate level and form of food aid, its financing, and how best to distribute it.

- reform of the National Food Corporation's program - eliminating mis-targetted subsidies, identifying which distribution measures reach those suffering food deficits, and how they should be expanded and financed.

- identification of a program of effective interventions - including evaluating the relative roles of food distribution, agricultural productivity measures, food storage, vulnerable group feeding, and promotion of effective practices (eg. weaning foods, feeding during pregnancy, etc.); followed by preparation of specific projects and agreement on financing for them.

8.12 Experience elsewhere shows that very targetted nutrition and feeding programs (eg. for malnourished children below the age of 36 months) can have a dramatic effect, although the capacity to administer them in Nepal is weak. No new feeding initiatives should be undertaken until the restructured Vulnerable Group Feeding program has operated for some time and is reviewed.

Poverty Alleviation Programs

8.13 As we have seen, the experience with programs to date has been discouraging. Transfer programs do not reach the poor, except for parts of NFC's remote area food distribution. Feeding programs have had only limited success, delivery is sporadic, and they are not well targetted. The food-for-work program is self-targetting, and while it has problems, is an area to consider expansion. The IRDP's show little sign of success in raising incomes of the poor; some intensive agricultural programs do, but their replicability is questionable (i.e. they may be successful because they are in the right places), and unit costs are high. Labour intensive construction and public works employment shows some promise, there are problems to be ironed out, but this is an area for expansion. The targetted credit programs, while partially successful, reach only a small proportion of the poor - their benefits appear to be more in group formation and other externalities than the provision of credit as such.

8.14   HMG should consider focusing its efforts on food-for-work and rural public employment schemes - not because they are necessarily the best solutions (ultimately their sustainability is questionable), but because they are the ones which can reach large enough numbers of the poor to make a difference, and which the administrative capacity exists to deliver. Many small income-generating programs (eg. under NGO's, IRDP's, and targetted credit projects) have had only very limited success. They require labour-intensive interventions, they have proven difficult to deliver, the capacity to expand them is limited by the shortage of skilled staff and by their high unit costs. HMG is probably better off: (i) concentrating its efforts on improving overall administrative and service delivery capacity; (ii) providing a fund for NGO's to continue to undertake small-scale income-generating projects; and (iii) pursuing a slow expansion of SFDP and PCRW - or a revised variant of them.

Credit Issues

8.15   HMG should exercise caution in pursuing major infusions of credit to fuel income growth among the poor - our analysis suggests that it will not lift binding constraints on the poor, and that the banking system can probably not efficiently deliver it. Certainly credit should form part of the Government's small business development strategy, but it should not be a centerpiece of poverty alleviation efforts. Providing small amounts of credit to the poor for selected investments is legitimate - for this HMG should focus on expansion of the SFDP and PCRW programs, but only after they have been consolidated and strengthened.

Education

8.16   HMG is following largely the right policies in education development - the capacity of the system needs to be expanded if it is to absorb the poor, and measures have to be found which will attract them into schools. HMG and the donor community need to be ready to allocate substantially more funds to general education (about double in real terms within the next 20 years). Analytical work should be undertaken on the determinants of school attendance, and to frame measures to increase enrollments. The following particular measures are recommended: (i) a national literacy campaign; (ii) revamping the curriculum and examination systems to improve relevance and quality; (iii) measures to increase female participation, including expansion of programs that allow alternative scheduling of classes, more female teachers, and possibly a scholarship program for girls.

Health

8.17   The greatest impact will come from non-health measures: water supply and hygiene, improved access to food, and reduced fertility. For the poor HMG should focus on a subset of simple health interventions -

immunization, oral rehydration, diarrheal disease management, and micronutrient supplementation - which do not rely on improvements in the health service for delivery.

8.18   A national mass hygiene campaign should be undertaken, which would combine nutrition education and awareness with a program of rural water supplies, and provision of sanitation facilities in both urban and rural areas.

General

8.19   There has been a startling failure to measure the income effects of massive expenditures on programs and projects designed with the intention of raising incomes (including those financed by the World Bank). In addition, there is no source which allows HMG to track overall changes in incomes or living standards. It is recommended that HMG establish the capacity to monitor household incomes on a periodic basis (the MPHBS is a good model), as well as requiring major projects to include specific income monitoring measures.

8.20   Given the critical importance of employment generation in off-farm activities surprisingly little is known about labour force and wage issues in Nepal. The Department of Labour focuses exclusively on the manufacturing sector (2% of employment). HMG should establish a data base and analytical capability covering employment statistics, wages and employment conditions. Complementary to this, a group of studies should be undertaken to improve the (almost non-existent) understanding of employment and incomes in the informal sector and in construction.

C.   Programming Implications

8.21   To promote equitable growth the Government needs to strengthen and expand key sectoral programs - including the national population campaign, modified approaches in agriculture, an expanded program of access, and the spread of basic education. In most of these areas steps are already underway - they need to be intensified, and implemented effectively. Major programming recommendations in these sectors are beyond the scope of this report, although additional suggestions are made in the text as to how they could have a greater impact on the reduction of poverty. In addition, to directly improve the conditions of the poor, the following six point action program is recommended as a framework around which HMG and donors could mobilize support.

(1)   Expansion of Selected Transfer Programs - probably consisting of a rural works program and Food-for-Work.

(2)   Incremental Improvements in Existing Income-Generating Projects - including the SFDP and PCRW, and a fund for NGO income-generating activities.

(3) **Food and Food Aid Programs** - including an agreed program of support, reform of the NFC, and identification of the most effective interventions.

(4) **Strengthening Non-Income Alleviation Measures** - which improve the welfare of the poor in the absence of income increases; including a mass hygiene campaign, simple preventive health interventions, and possibly targetted nutrition/feeding interventions.

(5) **A Package of Policy Reforms** - involving tenancy, labour contracting arrangements, service delivery and decentralization, and the regulatory framework for NGOs.

(6) **Analytical Work in Priority Areas** - including: (i) labour force issues, (ii) measuring income and income changes; (iii) development of migration and re-settlement strategies; (iv) tenancy and agrarian reform; (v) the construction industry; (vi) the informal sector; (vii) determinants of education participation; (viii) the private economics of family size; and (ix) the cost-effectiveness of simple health interventions.

8.22 In some of these areas further study is needed before program reforms can be finalized. The first steps in framing this program would be:

(a) preparation of sub-projects - including agreeing on modifications required to existing programs; appraisal of the new projects (eg. for food security and rural works); and agreement on the shape of the sectoral programs (population, agriculture, and rural access) - to the extent that they differ from, or represent an intensification of, on-going initiatives;

(b) in parallel, undertaking necessary analytical work (eg. for agrarian reform, food security measures); and,

(c) agreeing on a financing package for the main elements of the program.

ANNEX I
SUMMARY OF RECOMMENDATIONS

Annex I

SECTOR	RECOMMENDATIONS	REFERENCE
AGRICULTURE	- Continue to pursue small farmer irrigation programs in the hills and the terai; with careful design to improve benefits to the poor; consider using criteria under Irrigation Sector Program to select poor beneficiaries, where feasible.	Paras. 4.29, 6.63
	- For groundwater irrigation development in the terai: form small farmers into groups; regularize land tenure and registration; provide a package of credit and technical support.	Para. 4.29
	- Encourage development of a small traders to supply agricultural inputs - possibly by providing training and credit.	Para. 4.7, 4.30
	- Continue to pursue research and extention (R&E) based on a farming systems approach, including:	Paras. 4.11, 4.31
	(i) expansion of R&E based on systems free of purchased inputs;	
	(ii) conduct field tests using poor farmers as models;	
	(iii) more research on subsistence livestock; and,	
	(iv) consider a small-scale horticulture program for farmers in accessible areas.	Para. 4.19
	- Train agricultural extension workers to deal with female farmers	Para. 4.19
EMPLOYMENT	- Consider a major rural employment program, based around expansion of Food-for-Work, and SPWP.	Para. 6.69
	- In combination with above, consider a program of more labour-intensive construction works - as a first step do an analysis of costs and efficiency trade-offs.	Para. 4.44
	- Increase monitoring and supervision of construction contractors.	Para. 3.70
	- Enforce District wages.	Para. 3.69

Annex I

SECTOR	RECOMMENDATIONS	REFERENCE
	- Review construction contracting and employment practices, including scope for use of smaller contractors.	Para. 4.42
	- Increase labour-intensive maintenance, especially on rural roads.	Para. 4.43
	- Rather than beneficiary operations and maintenance (eg. of irrigation schemes), consider a user charge combined with wage employment of poorer villagers.	Para. 6.63
LAND TENURE	- De-link tenancy reform from land redistribution issues.	Para. 3.42
	- Design and implement a tenancy and agrarian reform package.	Para. 3.43
	- As a first step, analyze the extent and type of tenancies, and their economic effects.	Para. 3.43
	- Streamline procedures for land registration and transfer.	Para. 3.42
	- Divorce issues of urban land management from those of agrarian reform.	Para. 3.41
	- Separate food distribution for civil servants from NFC's other activities.	Para. 6.45
FOOD SECURITY	- Preferably discontinue distribution to civil servants and replace with salary increases.	Paras. 6.21, 6.45
	- Drop Kathmandu entirely from NFC's operations.	Para. 6.45
	- Review and determine scope for benefit targetting in remote areas under NFC.	Para. 6.45
	- Undertake no new feeding programs at this stage.	Para. 6.47
	- Review experience with the Vulnerable Group Feeding program next year, along with experience elsewhere.	Para. 6.47
	- HMG and donors should undertake a general assessment of options for food security interventions, and agree on a program.	Para. 8.11

Annex I

SECTOR	RECOMMENDATIONS	REFERENCE
POPULATION	- Mount a major national population campaign, it should include:	Para. 5.52
	Recruitment of a large number of female family planning workers (maybe 25,000), highly paid and well-trained.	
	Population linked incentives in other sectors.	
	Possibly a separate service delivery system, outside the health service.	
	- Develop a cadre of population specialists and managers.	Par. 5.52
	- Undertake analytical work on the income effects and determinants of of family size.	Para. 5.54
EDUCATION	- Concentrate on basic education and literacy	Para. 5.43
	- Be prepared to increase financing for primary education from US$24 million to US$50 million p.a.	Para. 5.44
	- Undertake a national literacy campaign.	Para. 5.45
	- Improve curriculum relevance, reduce rote-learning and dependence on examinations.	Para. 5.46
	- Consider automatic promotion in the first three years, and possibly instruction in non-Nepali languages.	Para. 5.46
	- Consider (i) siting schools closer to families; (ii) more flexible hours, (iii) recruiting teachers locally.	Para. 5.46
	- Increase the use of female teachers (postpone SLC requirement), consider providing female scholarships and uniforms.	Para. 5.47
HEALTH	- Focus on immunization, oral rehydration, micro-nutrient programs.	Para. 5.60

Annex I

SECTOR	RECOMMENDATIONS	REFERENCE
	- Consider a mass national hygiene campaign – combining hygiene education, and changing hygiene behavior, with rural water supplies and sanitation facilities.	Paras. 6.48, 5.59
	- In conjunction with the above – a nutrition behavior program to change weaning and feeding practices.	Para. 6.48
CREDIT	- Consolidate SFPD and PCRW before further expansion.	Para. 6.54
	- Strengthen the training of group organizers.	Para. 6.54
	- Divorce development functions from commercial banking activities.	Para. 6.57
	- Consider establishment of a "Poverty Bank" to finance small amounts of directed credit for income increasing activities among the poor.	Para. 6.57
	- Reconsider massive expansions of commercial credit to the poor.	Paras. 6.60, 8.15
SPATIAL ISSUES	- Undertake an analysis of measures needed to manage population transition out of the hills – including scope for conversion of terai forests.	Para. 3.22
	- Consider concentrating on growth of small-to-medium sized settlements.	Para. 7.15
	- Adopt a hierarchy of support for hills areas, depending on degree of accessibility and sustainability of welfare support and economic activity.	Paras. 8.09, 7.15
GENERAL	- Consider a land tax – except on small landowners.	Para. 6.05
	- Strengthen general programs, and provide infrastructure, rather than finance additional IRDP's.	Para. 6.32
	- If there are to be further IRDP's, they should focus on very few activities – probably roads and irrigation.	Para. 6.32

Annex I

SECTOR	RECOMMENDATIONS	REFERENCE
	- Enforce collection of irrigation charges.	Para. 6.19
	- Replace irrigation capital subsidies with a credit program, with an exemption for very small landholders.	Para. 6.20
	- Undertake analytical work to identify possible micro interventions to accelerate informal sector income growth.	Para. 8.06
	- Undertake a systematic review of administration and personnel factors effecting service-delivery, rather than tackling them sector-by-sector.	Para. 6.77
	- Pursue programs which put more income in the hands of women (eg. PCRW).	Para. 3.91
	- Expand programs which reduce the time spent (mostly by women) in gathering fuel fodder, and water - eg. rural water supply, community forestry, trail improvement.	
	- Establish a data-base on employment and wages issues.	Para. 8.20
	- Improve monitoring of incomes, income changes, and the income effects of projects and programs.	Para. 8.19
	- Consider partnerships of NGOs executing some projects financed by large donors.	Para. 6.84
	- Consider a fund for small income-generating projects to be implemented by NGO's, along with technical support.	Para. 6.73
	- Review the regulatory framework for NGO's, with a view to relaxing it.	Para. 6.86
	- Build up the capacity of local NGO's, consider a small fund through which INGO's and others could support them.	Para. 6.86

ANNEX II

STATISTICAL APPENDICES

Annex II - Statistical Appendices

		Page
Annex II.1	Country Data..	139
1.1	Key Country Data.......................................	141
1.2	Changes in Key Economic Variables......................	142
Annex II.2	MPHBS Special Tabulations...............................	143
2.0	Description of the MPHBS...............................	145
	Figure 1 - Average Incomes by Decile...................	146
2.1	Characteristics of the Poor............................	147
2.2	Composition of Incomes.................................	148
2.3	Consumption Patterns...................................	150
2.4	Composition and Sources of Food Consumption............	152
2.5	Time use by Age, Gender, and Income....................	156
2.6	Education Participation by Income......................	157
Annex II.3	Labour Force Data.......................................	159
3.1	Occupational Classification by Incomes.................	161
3.2	Utilization of Working Days............................	163
3.3	Sample Distribution of Workdays........................	165
3.4	Nepal Labour Force - 1981 and 1989.....................	166
3.5	Estimated Sectoral Breakdown of the Labour Force.......	166
3.6	Some Selected Wage Rates...............................	167
3.7	Sample Estimates of Remittances by Migrants............	167
Annex II.4	Formal Sector..	169
4.1	Manufacturing Sector Output and Employment Growth......	171
4.2	Formal Sector Employment (1990) and Projections........	172
4.3	Manufacturing Employment and Output 1986/87............	173
4.4	Cost Composition of Earthworks Contracts...............	174
4.5	Estimated Financial Flows Under Infrastructure Projects...................................	175
Annex II.5	The Formal Sector......................................	177
5.1	Informal Sector Employment.............................	179
5.2	Incidence of Poverty Amongst Informal Sector Workers...	180
5.3	Estimated Household Informal Sector Incomes............	181
5.4	Informal Sector Employment Projections.................	182
Annex II.6	Agriculture and Land Tenure............................	183
6.1	Indexes of Cultivated Area, Inputs, Production, and Yield...	185
6.2	Incidence of Irrigation by Farm Size...................	186
6.3	Characteristics of Farmer Groups in the Hills..........	187
6.4	Yield Assumptions for Farm Models......................	188
6.5	Potential Income Per Hectare from Field Crops in the Hills..	189

6.6	Minimum Land Holding Necessary to Earn a Poverty Line Income from Field Crops in the Hills..........	189
6.7	Estimated Household Income to be Derived from Horticulture..	190
6.8	Household Incomes and Landholdings in the Terai by Level of Poverty...........................	191
6.9	Potential Income Per Hectare from Field Crops in the Terai..	191
6.10	Estimates of Total Labour Use in Agriculture...........	192
6.11	Estimates of Percentage of Landless Households........	196

Annex II.7 **Credit and Debt**.. 197

7.1	Proportion of Farm Families Borrowing from Different Sources by Farm Size.....................	199
7.2	Comparative Data on Average Borrowings by Region.......	200
7.3	Source of Loan by Ecological Region...................	200
7.4	Purpose of Borrowing for Small Farm Families...........	201
7.5	Purpose of Borrowing in Sampled Terai Village by Economics Status, 1984...........................	201

Annex II.8 **Nutrition and Food Security**............................. 203

8.1	Share of Cereals in Food Consumption by Income........	205
8.2	Year-to-Year Variation in Food Production.............	206
8.3	Average Cereal Prices by Region.......................	207
8.4	Summary of District Food Balances.....................	208
8.5	Prevalence of Malnutrition............................	209
8.6	Nutritional Status of Children........................	210
8.7	NFC Food Distribution and Transportation Costs.........	211

ANNEX II.1 - COUNTRY DATA

Annex II.1

Table 1.1

Key Country Data /a

	Population	Consumer Price Index	Exchange Rate		Real GDP	Foodgrain Production
	(millions)	(1972/73 = 100)	Rs/US$	Rs/IRs	(1974/75 Rs. blns.)	(000 mt.)
1950	8.3				9.4	3469
1955	8.7		7.7		10.8	3537
1960	9.4		7.6		12.1	3748
1965	10.4		7.6	1.60	13.5	3335
1970/71	11.6		10.1	1.35	15.4	3557
1975/76	13.2	137	12.3	1.39	17.3	3908
1980/81	15.0	201	11.9	1.45	20.2	3829
1981/82	15.4	222	13.2	1.45	20.9	3983
1982/83	15.8	254	13.9	1.45	20.3	3350
1983/84	16.2	269	15.3	1.45	22.3	4289
1984/85	16.5	281	17.8	1.45	23.6	4211
1985/86	16.9	326	19.8	1.45	24.6	4437
1986/87	17.4	368	21.5	1.68	25.3	4094
1987/88	17.8	409	22.1	1.68	27.8	4749
1988/89	18.2	438	25.0	1.68	28.2	5395
1989/90	18.8		28.0	1.68		

a/ NOTE: These are the price index, population, and GDP series which underlie calculations throughout the report.

Annex II.1

Table 1.2

Changes in Key Economic Variables

	Average Annual Growth Rates		
	1965-73	1973-80	1980-87
GDP	1.7%	3.0%	4.4%
Agriculture	1.5%	0	3.9%
Foodgrain Production	1.0%	-0.7%	2.4%
Inflation (Consumer Prices)	5.3%	6.9%	10.3%

ANNEX II.2 - MPHBS SPECIAL TABULATIONS

Annex II.2

Table 2.0: The Multi-Purpose Household Budget Survey

The Multi-Purpose Household Budget Survey (MPHBS) was conducted by the Nepal Rastra Bank in 1984/85. The study includes modules on household composition, education levels, housing, employment, income, expenditures and time use. Special attention was paid to non-monetized transactions, which were valued at prices obtaining in the local markets adjacent to the household in question.

The survey used a cluster sampling technique to select 23 districts and twelve town panchayats. Five inaccessible mountain districts were excluded from the sampling frame. Survey teams enumerated all households (using the United Nations Statistical Office definition of household), in the selected areas, excluding 'beggars' and people living in institutions such as schools and prisons. In total, 3,662 households were interviewed.

Households chosen in the sample were given two sets of interviews, one in the rainy season and one in the dry season, in order to provide a balanced picture of consumption and expenditure patterns. During each set of interviews households were asked demographic and housing information, as well as expenditure, income and savings patterns for the previous month. Part of each set of interviews included daily visits by enumerators for seven consecutive days to collect 24 hour recall data on purchases of food, use of home produced food, food received free and food received as wages.

Field supervisors reinterviewed ten percent of the households to check on the quality of the interviews. In addition, internal checks were made such as comparing sources of income in the previous month to time allocation in the previous month and comparing income to expenditure. In the latter case, if expenditures exceeded income by more than fifteen percent, the household was reinterviewed to insure the validity of the information.

These data represent the best available national data on incomes and expenditure in Nepal. The survey was well organized and implemented and the data collection carefully monitored. As with many such surveys the very wealthy and the very poor are probably underrepresented. The exclusion of "beggars" (no definition given) means that a significant number of the very poor, particularly in urban areas, may have been missed. On the whole, however, the surveys presents an accurate picture of incomes in Nepal.

ANNEX II.2

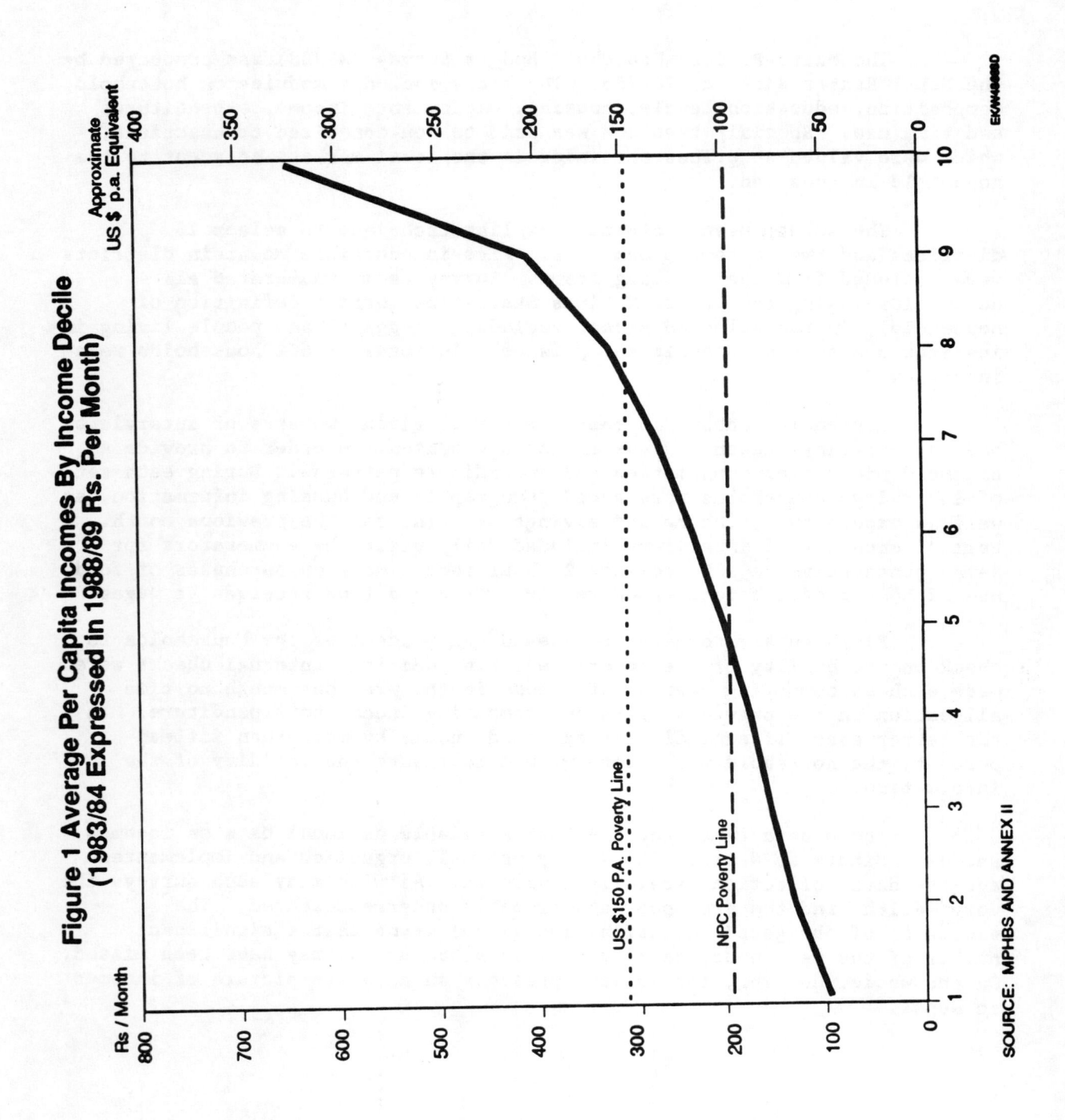

Figure 1 Average Per Capita Incomes By Income Decile (1983/84 Expressed in 1988/89 Rs. Per Month)

SOURCE: MPHBS AND ANNEX II

NEPAL

POVERTY AND INCOMES STUDY

TABLE 2.1: CHARACTERISTICS OF THE POOR

(1984 Rs. per month)

| | RURAL ||||||| URBAN |||||
| | Terai || Hills || Mountains || Terai || Hills ||
	Poor	Non-Poor	Poor	Non-Poor	Poor	Non-Poor	Poor	Non-Poor	Poor	Non-Poor
Average Household Size	7	7	6	5	6	4	7	6	6	5
Average Monthly Income	702	1,493	673	1,452	886	1,186	704	1,379	815	1,919
Average Monthly Expenditure	741	1,247	738	1,273	869	1,055	765	1,118	971	1,645
Per Capita Monthly Income	99	221	111	267	139	271	102	249	131	377
No. of Earners Per Household	3	3	3	3	4	3	2	2	3	2
Dependency Ratio	1.3	1.1	0.9	0.8	0.8	0.6	1.7	1.7	2.6	1.7
No. of Persons/Sleeping Room	3.8	3.0	3.8	2.8	3.5	2.6	3.6	2.8	3.5	2.4
Literacy Rate (%)	22.0%	40.2%	37.1%	51.1%	32.1%	42.1%	35.2%	59.0%	49.1%	72.4%
Enrollment Rate (%)										
Primary	30.1%	53.0%	49.2%	64.5%	37.1%	55.1%	37.2%	63.4%	64.9%	79.0%
Secondary	13.0%	29.7%	10.9%	29.8%	17.8%	21.3%	22.0%	42.0%	23.9%	48.5%

(Source: MPHBS 1989)

Annex II.2

Annex II.2

NEPAL

POVERTY AND INCOMES STUDY

TABLE 2.2A: COMPOSITION OF INCOME BY TYPE

(1984 Rs. per month)

	RURAL							URBAN			
	Terai		Hills		Mountains		Terai		Hills	
	Poor	Non-Poor	Poor	Non-Poor	Poor	Non-Poor	Poor	Non-Poor	Poor	Non-Poor
Cash Income										
Wages & Salaries	108	107	121	209	123	173	301	417	350	721
Agriculture	64	398	56	262	112	199	26	90	52	83
Other Family Enterprises	29	77	29	115	18	58	105	372	76	460
Other (Prop., Interest, etc.)	24	64	30	94	19	58	30	143	45	339
Subtotal	224	646	236	679	272	487	461	1,023	523	1,603
Income In Kind										
Food & Own Production	33	70	33	58	52	63	21	34	23	31
As Part of Wages	120	52	39	19	39	21	34	21	12	19
Family Enterprises	286	688	307	635	456	552	162	276	197	231
Imputed/Received Free	38	36	58	61	66	62	26	25	60	35
Subtotal	478	847	437	773	613	698	244	357	292	315
TOTAL	702	1,493	673	1,452	885	1,186	705	1,379	815	1,919
Rental Value of own home	41	72	38	87	75	79	98	179	126	356

Notes: * Almost all in the form of food; includes agricultural production and rental income in kind (i.e. sharecropping payments).
** Includes rental value of rent-free dwelling.

NEPAL

POVERTY AND INCOMES STUDY

TABLE 2.2B: COMPOSITION OF INCOMES BY SOURCE

(1984 Rs. per month)

| | RURAL ||||||| URBAN ||||||
| | Terai || Hills || Mountains || Terai || Hills ||
	Poor	Non-Poor	Poor	Non-Poor	Poor	Non-Poor	Poor	Non-Poor	Poor	Non-Poor
Agricultural Income										
Cash	64	398	56	262	112	199	26	90	52	83
Kind	285	684	306	628	456	549	160	271	197	220
Subtotal	349	1,082	363	890	568	748	186	361	248	303
Non-Agricultural Enterprises										
Cash	29	77	29	115	18	58	105	372	76	460
Kind	1	4	1	7	2	3	2	5	1	11
Subtotal	29	80	30	121	20	61	106	377	77	472
Wages & Salaries										
Cash	108	107	121	209	123	173	301	417	350	721
Kind	120	52	39	19	39	21	34	21	12	19
Subtotal	228	160	159	228	162	193	335	438	362	740
Other Cash Income										
Rental	5	19	2	7	1	17	9	59	24	167
Others	19	44	28	88	17	41	21	84	22	172
Other Income in Kind										
Home Produced	33	70	33	58	52	63	21	34	23	31
Received Free	38	36	58	61	66	62	26	25	60	35
TOTAL	702	1,493	673	1,452	885	1,186	705	1,379	815	1,919
Rental Value of own home	41	72	38	87	75	79	98	179	126	356

Annex II.2

… - 150 - Annex II.2

NEPAL

POVERTY AND INCOMES STUDY

TABLE 2.3A: HOUSEHOLD CONSUMPTION PATTERN

(1984 Rs. per month)

| | RURAL ||||||| URBAN ||||
| | Terai || Hills || Mountains || Terai || Hills ||
	Poor	Non-Poor	Poor	Non-Poor	Poor	Non-Poor	Poor	Non-Poor	Poor	Non-Poor
Family Size	7	7	6	5	6	4	7	6	6	5
Consumption										
Grains and Pulses	409	524	355	475	434	411	388	382	411	453
Oils, Spices, etc.	37	58	37	58	45	50	42	57	56	79
Vegetables & Fruits	49	67	51	71	51	58	58	73	74	121
Meat, Fish & Eggs	15	30	19	38	24	30	17	31	31	65
Other Foodstuffs	37	78	44	113	56	75	44	120	70	185
Fuel, Light, Water, etc.	51	72	57	74	69	67	49	64	85	119
Clothing	56	170	78	183	86	159	64	138	87	205
Education & Health	31	107	31	82	22	37	43	104	60	151
Transport	6	23	5	15	4	11	8	26	10	51
Religious Obligations	2	12	4	13	3	9	1	6	2	13
All Other	47	106	69	152	76	149	69	116	79	205
Total Consumption Expenditure	741	1,247	749	1,273	869	1,055	783	1,118	965	1,645
Non-Consumption Expenditure	36	111	12	54	11	45	3	38	64	52
Rental Value of Own Home	41	72	38	87	75	79	98	179	126	356

NEPAL

POVERTY AND INCOMES STUDY

TABLE 2.3B: PER CAPITA CONSUMPTION PATTERN

(1984 Rs. per month)

| | RURAL ||||||| URBAN ||||
| | Terai || Hills || Mountains || Terai || Hills ||
	Poor	Non-Poor	Poor	Non-Poor	Poor	Non-Poor	Poor	Non-Poor	Poor	Non-Poor
Family Size	7	7	6	5	6	4	7	6	6	5
Consumption										
Grains and Pulses	58	74	50	67	61	58	55	54	58	64
Oils, Spices, etc.	5	8	5	8	6	7	6	8	8	11
Vegetables & Fruits	7	9	7	10	7	8	8	10	10	17
Meat, Fish & Eggs	2	4	3	5	3	4	2	4	4	9
Other Foodstuffs	5	11	6	16	8	11	6	17	10	26
Fuel, Light, Water, etc.	7	10	8	10	10	9	7	9	12	17
Clothing	8	24	11	26	12	22	9	19	12	29
Education & Health	4	15	4	11	3	5	6	15	8	21
Transport	1	3	1	2	1	2	1	4	1	7
Religious Obligations	0	2	1	2	0	1	0	1	0	2
All Other	7	15	10	21	11	21	10	16	11	29
Total Consumption Expenditure	104	176	106	179	122	149	110	157	136	232
Non-Consumption Expenditure	5	16	2	8	2	6	0	5	9	7

Annex II.2

NEPAL

POVERTY AND INCOMES STUDY

TABLE 2.4A: COMPOSITION & SOURCES OF FOOD CONSUMPTION
(In K.Cal/person/day)

Rural Terai

	Poor			Non-Poor		
	Market Purchased	Home Produced *	Bartered	Total	Market Purchased	Home Produced *	Bartered	Total
Grains, Cereals & Pulses	215	1,208	400	1,823	139	1,983	163	2,285
Oils, Fats & Spices	25	13	1	38	65	32	0.4	97
Fruit & Vegetables	35	45	0.6	81	34	69	1	104
Meat, Fish & Eggs	4	5	0.1	9	9	6	0	15
All Other Foods	31	43	2	76	90	93	0	183
TOTAL	309	1,314	403	2,027	337	2,182	165	2,684

* Includes Received Free.

(Source: MPHBS 1989)

Annex II.2

Annex II.2

NEPAL

POVERTY AND INCOMES STUDY

TABLE 2.4B: COMPOSITION & SOURCES OF FOOD CONSUMPTION
(In K.Cal/person/day)

Urban Terai

	Poor				Non-Poor		
	Market Purchased	Home Produced *	Bartered	Total	Market Purchased	Home Produced *	Bartered	Total
Grains, Cereals & Pulses	907	818	115	1,840	881	975	64	1,920
Oils, Fats & Spices	41	9	1	50	70	35	0.4	105
Fruit & Vegetables	54	20	0	74	81	23	0.2	104
Meat, Fish & Eggs	5	2	0	7	11	2	0.0	12
All Other Foods	39	21	(0)	60	118	40	0.0	157
TOTAL	1,046	869	116	2,031	1,160	1,073	65	2,298

* Includes Received Free.

(Source: MPHBS 1989)

NEPAL

POVERTY AND INCOMES STUDY

TABLE 2.4C: COMPOSITION & SOURCES OF FOOD CONSUMPTION
(In K.Cal/person/day)

Rural Hills

	Poor				Non-Poor		
	Market Purchased	Home Produced *	Bartered	Total	Market Purchased	Home Produced *	Bartered	Total
Grains, Cereals & Pulses	366	1,322	135	1,823	272	1,807	43	2,122
Oils, Fats & Spices	36	24	0	60	32	43	0.2	76
Fruit & Vegetables	10	56	0.5	67	25	85	1	111
Meat, Fish & Eggs	5	3	0.0	8	12	5	0.0	17
All Other Foods	62	165	4	231	115	247	3	365
TOTAL	480	1,569	139	2,189	456	2,188	48	2,691

* Includes Received Free.

(Source: MPHBS 1989)

Annex II.2

NEPAL

POVERTY AND INCOMES STUDY

TABLE 2.4D: COMPOSITION & SOURCES OF FOOD CONSUMPTION
(In K.Cal/person/day)

Urban Hills

	Poor				Non-Poor			
	Market Purchased	Home Produced *	Bartered	Total	Market Purchased	Home Produced *	Bartered	Total
Grains, Cereals & Pulses	976	766	19	1,761	1,124	607	30	1,762
Oils, Fats & Spices	55	11	0.0	66	105	17	0.2	122
Fruit & Vegetables	51	29	0.0	80	107	29	0.4	136
Meat, Fish & Eggs	12	0.5	0.0	12	28	2	0.0	30
All Other Foods	94	77	0.0	172	253	107	0.2	361
TOTAL	1,188	883	19	2,090	1,617	762	31	2,410

* Includes Received Free.

(Source: MPHBS 1989)

Annex II.2

Annex II.2

Table 2.5 - Time Use by Age, Gender, and Income

Time Use - Average Hours Per Day

	Rural				Urban			
	Terai		Hills		Terai		Hills	
	Poor	Non-Poor	Poor	Non-Poor	Poor	Non-Poor	Poor	Non-Poor
Men > 15 years old								
Conventional Economic Activities	5.33	4.88	4.29	4.18	6.40	5.76	5.17	5.17
Subsistence Economic Activities	0.97	0.92	1.67	1.73	0.39	0.32	0.87	0.34
Domestic Work	1.67	1.75	1.95	2.13	1.64	1.56	1.51	1.39
Work Burden	7.97	7.55	7.91	8.04	8.43	7.64	7.55	6.90
Education and Reading	0.67	0.98	0.84	1.05	0.72	1.64	1.33	2.68
Women > 15 years old								
Conventional Economic Activities	1.71	1.50	2.64	2.62	1.36	1.30	2.67	1.79
Subsistence Economic Activities	1.91	1.68	2.58	2.42	1.38	1.01	1.97	1.18
Domestic Work	5.94	5.90	5.24	5.95	6.71	6.65	5.26	5.65
Work Burden	9.56	9.08	10.46	10.99	9.45	8.96	9.90	8.62
Education and Reading	0.26	0.42	0.15	0.25	0.45	0.59	0.46	1.55
Boys 10 to 14 years old								
Conventional Economic Activities	1.45	0.95	1.05	1.27	1.20	1.17	0.51	0.33
Subsistence Economic Activities	0.72	0.56	1.18	1.19	0.40	0.27	0.78	0.44
Domestic Work	1.61	1.33	1.55	1.95	1.07	0.79	0.82	0.76
Work Burden	3.78	2.84	3.78	4.41	2.67	2.23	2.11	1.53
Education and Reading	2.85	4.47	3.52	3.84	4.19	4.96	5.07	6.31
Girls 10 to 14 years old								
Conventional Economic Activities	1.39	0.91	1.91	1.75	0.60	0.45	1.11	0.63
Subsistence Economic Activities	1.52	1.32	2.18	1.96	0.89	0.76	1.41	0.82
Domestic Work	3.62	3.14	2.73	3.95	4.18	3.16	3.08	2.18
Work Burden	6.53	5.37	6.82	7.66	5.67	4.37	5.60	3.63
Education and Reading	1.03	2.37	1.44	2.05	2.00	3.67	2.73	5.18
Boys 6-9 years old								
Work Burden	1.30	1.34	1.90	1.70	1.78	0.79	0.83	0.44
Education and Reading	2.26	3.08	3.22	4.08	2.93	4.79	3.66	6.33
Girls 6-9 years old								
Work Burden	3.12	2.21	3.26	3.19	2.84	1.99	2.58	1.20
Education and Reading	1.03	2.30	1.55	2.08	1.46	3.61	1.71	5.51

Source: MPHBS Special Tabulations.

Annex II.2

TABLE 2.6: EDUCATION ENROLLMENTS AMONG THE POOR

	Poor		Non-Poor	
	Primar	Secondary	Primar	Secondary
Rural Hills				
Male	72%	21%	79%	41%
Female	33%	6%	54%	18%
Total	53%	12%	67%	30%
Rural Terai				
Male	45%	20%	67%	39%
Female	14%	5%	39%	18%
Total	30%	13%	53%	30%
Urban Hills				
Male	75%	38%	83%	58%
Female	50%	8%	75%	39%
Total	64%	25%	79%	49%
Urban Terai				
Male	49%	32%	72%	48%
Female	26%	7%	51%	35%
Total	38%	22%	63%	42%

(Source: MPHBS 1989)

ANNEX II.3 - LABOR FORCE DATA

Annex II.3

TABLE 3.1A: OCCUPATIONAL CLASSIFICATION BY INCOME

RURAL

	Terai Poor	Terai Non-Poor	Hills Poor	Hills Non-Poor
Professional, Technical Administrative & Managerial	0.9%	1.7%	0.9%	2.3%
Office Workers	0.4%	1.0%	0.9%	2.7%
Sales and Service	4.8%	13.2%	2.0%	8.1%
Agricultural Workers	81.6%	78.7%	80.5%	75.9%
(of which: Farm Labourers)	27.3%	11.9%	6.4%	2.2%
Production Workers	3.3%	1.8%	4.4%	3.9%
Construction, Transport & Communications	8.9%	3.6%	11.5%	7.1%
(of which: Casual Labourers)	8.2%	3.1%	10.7%	6.1%

(Source: MPHBS 1989)

Annex II.3

TABLE 3.1B: OCCUPATIONAL CLASSIFICATION BY INCOME

URBAN

	Terai Poor	Terai Non-Poor	Hills Poor	Hills Non-Poor
Professional, Technical Administrative & Managerial	0.3%	5.3%	–	9.3%
Office Workers	3.2%	9.2%	5.1%	13.0%
Sales and Service	17.1%	26.4%	4.3%	22.7%
Agricultural Workers	46.6%	34.4%	50.5%	29.4%
(of which: Farm Labourers)	5.0%	2.6%	2.9%	0.4%
Production Workers	8.9%	11.7%	9.6%	14.4%
Construction, Transport & Communications	24.3%	13.5%	32.4%	10.8%
(of which: Casual Labourers)	22.4%	9.4%	30.2%	6.9%

(Source: MPHBS 1989)

TABLE 3.2A: UTILIZATION OF WORKING DAYS

RURAL NEPAL

	TERAI						HILLS						MOUNTAINS					
	Poor			Non-Poor			Poor			Non-Poor			Poor			Non-Poor		
	Male	Female	Total	Male	Female	Total	Male	Female	Total	Male	Female	Total	Male	Female	Total	Male	Female	Total
Total Days Available	501	355	856	536	328	864	368	422	790	352	420	772	423	434	857	334	362	696
Self-Employment	132	72	204	229	97	326	106	133	239	140	153	293	142	140	282	154	166	320
Labor Exchange	2	2	4	3	2	5	9	16	25	12	25	37	26	33	59	18	30	48
Off-Farm Employment	174	95	269	77	43	120	85	53	137	49	30	79	83	38	121	50	23	73
Unutilized	193	185	379	227	186	413	165	224	389	151	212	363	173	222	395	112	143	255

(Source: MPHBS 1989)

Annex II.3

TABLE 3.2B: UTILIZATION OF WORKING DAYS

URBAN NEPAL

	TERAI Poor Male	TERAI Poor Female	TERAI Poor Total	TERAI Non-Poor Male	TERAI Non-Poor Female	TERAI Non-Poor Total	HILLS Poor Male	HILLS Poor Female	HILLS Poor Total	HILLS Non-Poor Male	HILLS Non-Poor Female	HILLS Non-Poor Total
Total Days Available	398	238	636	362	197	559	289	329	618	262	256	518
Self-Employment	136	53	189	176	69	245	72	78	149	130	99	229
Labor Exchange	0	0	0	1	-	1	8	13	21	4	7	11
Off-Farm Employment	177	56	232	100	26	126	118	118	237	68	40	108
Unutilized	85	130	215	85	102	187	91	119	211	60	110	170

TABLE 3.3: SAMPLE SURVEY DISTRIBUTION OF WORKDAYS

(Days Per Person)

	Percentage of Workers	Self-Employment	(Of Which Non-Farm)	Wage Employment	(Of Which Non-Farm)	Not Classified	Total	Non-Farm Wage Employment in Classified Total
Landless	35.3%	62	(25)	62	(33)	55	179	27%
0.0 - 0.5 h.a.	18.1%	80	(13)	33	(20)	33	146	18%
0.5 - 1.0 h.a.	11.8%	101	(14)	47	(24)	22	170	16%
1.0 - 2.0 h.a.	13.6%	90	(10)	35	(16)	21	146	13%
2.0 - 10.0 h.a.	20.2%	72	(7)	14	(6)	42	128	7%
10.0 h.a. +	1.0%	100	(5)	10	(6)	98	208	5%
All	100.0%	76		41		40	157	

Source:

Acharya <u>A Study of Rural Labour Markets in Nepal</u> - 1987.

Annex II.3

Table 3.4: Nepal Labour Force 1981 and 1989
(millions)

	1981 Male	1981 Female	1981 Total	1989 (est.) Male	1989 (est.) Female	1989 (est.) Total
Population	7.7	7.3	15.0	9.4	9.0	18.4
Economically Active (over age 10)	4.5	2.4	6.9	5.7	3.4	9.1
Estimated LF - Participation Rates: (aged 10 and over)	83%	46%	65%	83%	57%	70%

Source: 1981 Census; 1989 mission estimates.

Table 3.5: Estimated Sectoral Breakdown of the Labour Force

Agriculture	86%
Manufacturing	2%
Construction, Utilities, Transport	2%
Commerce	3%
Services	5%
Others	2%

Source: Mission estimates from various sources.

Annex II.3

Table 3.6: Some Selected Wage Rates (1988)

	Rs./day
Agricultural Wage Labour:	Rs. 10-15
Construction & Public Works:	Rs. 15-30
Cottage/Small Industries:	Rs. 15-30
Agricultural Processing:	Rs. 25-50
General Portering:	Rs. 20-50

Source: ERL - from Interviews

Table 3.7: Sample Estimates of Remittances by Migrants

	Remittances as a Proportion of Income	Average Remittances
Low Income	2-5%	Rs. 120-210
Medium Income	5-11%	Rs. 340-680
High Income	18-24%	Rs. 1190-2490

SOURCE: Report on *Rasuwa-Nuwakot Rural Development Program* (1982) as reported in New Era, *op. cit.*.

ANNEX II.4 - FORMAL SECTOR

Annex II.4

Table 4.1

The Manufacturing Sector - Output & Employment Growth

(Output in Rs. millions, current prices)

	1976/77 Gross Output	1976/77 Employment	1981/82 Gross Output	1981/82 Employment	1986/87 Gross Output	1986/87 Employment	Avg Annual Empl Growth 1976/77-1981/82	Avg Annual Empl Growth 1981/82-1986/87	Avg Annual Empl Growth 1976/77-1986/87
Food	3394	20835	4185	25463	4940	32124	4%	5%	5%
Drink	134	6887	987	11087	363	11195	10%	-	5%
Textiles	174	8323	655	13911	2154	35575	11%	20%	16%
Wood/Paper	144	5970	612	7267	605	11829	4%	10%	8%
Plastics,etc.	16	740	195	830	863	7360	3%	55%	25%
Non-Metallic	59	7550	130	16640	938	45490	16%	22%	18%
Metallic	46	1390	192	3843	771	5810	23%	9%	15%
Total (incl. misc. others)	4238	59037	7062	80150	13537	152579	6%	14%	10%

Annex II.4

Table 4.2

Formal Sector Employment & Projections

		Low Scenario		Middle		High Scenario	
Sector	1990 Employment	Growth Rate	2010 Employment	Growth Rate	2010 Employment	Growth Rate	2010 Employment
Manufacturing	150,000	4.5%	361,757	6.5%	528,547	10.0%	1,009,125
Construction	350,000	1.0%	427,067	3.0%	632,139	5.0%	928,654
Tourism	10,000	4.0%	21,911	6.5%	35,236	9.0%	56,044
Public Sector	130,000	2.0%	193,173	3.0%	234,794	4.0%	284,846
Trade	60,000	4.5%	144,703	6.5%	211,419	9.0%	336,265
Transport	18,000	2.5%	29,495	3.5%	35,816	6.0%	57,728
Services	100,000	5.0%	265,330	7.0%	386,968	10.0%	672,750
Total Formal Sector:							
w/ Construction	818,000		1,443,436		2,064,920		3,345,412
w/out Constructi	468,000		1,016,369		1,432,781		2,416,758

2010 Employment (thousands)

	Low	Middle	High
- w/construction	1,443	2,065	2,857
Share of LF:	10.6%	15.2%	21.0%
- w/out	1,016	1,433	1,928
Share of LF:	7.5%	10.5%	14.2%

2010 - New Jobs Being Created Annually

	Low	Middle	High
- w/construction	45,805	104,737	194,002
Share of LF Growth:	12.0%	27.5%	50.9%
- w/out	41,534	85,773	147,570
Share of LF Growth:	10.9%	22.5%	38.7%

Table 4.3: Manufacturing Employment and Output - 1986/87
(Employment in numbers, values in current Rs. millions)

Industry	Employment	Gross Output	Value Added
Structural Clay	41,492	456	275
Grain Mills	21,890	5,295	945
Other Agricultural Processing	12,714	795	360
Textiles and Garments	17,570	860	380
Jute, Spinning and Weaving	15,730	1,000	390
Sawmills and Furniture	6,740	335	130
Printing and Allied	3,400	153	70
Metal Fabrication, etc.	3,250	587	160
Total (includes misc. others):	152,580	13,540	4,490

Source: CBS, Census of Manufacturers.

Table 4.4: Cost Composition of Earthworks Contracts

	Project Area	Kathmandu	Abroad	
Salaries and Wages	33%	8%	-	41%
Equipment and Tools	1%	-	5%	6%
Taxes and Interests	-	6%	-	6%
Overheads	3%	12%	-	15%
Contractors Profit	1%	16%	3%	20%
Sub-Contractors Profits	9%	2%	-	11%
Total	47%	44%	8%	100%

Earthworks Contract - Piece Work

	Project Area	Kathmandu	Abroad	
Salaries and Wages	41%	8%	-	49%
Equipment and Materials	-	2%	30%	32%
Taxes and Interests	2%	9%	-	11%
Overheads	2%	1%	-	3%
Contractors Profits	5%	-	-	5%
Total	50%	20%	30%	100%

Source: From Lamosung - Jiri Road project.

Table 4.5: Estimated Financial Flows Under Infrastructure Projects

	Within Project Area	Kathmandu or Other Urban Area	Outside Nepal
Mountain Road	32%	27%	41%
Suspension Bridge	35%	45%	31%
Drinking Water	30%	20%	50%
Irrigation			
- Large Scale	32%	27%	41%
- Small Scale	60%	20%	20%
- Community Type	90%	5%	5%
Trail Construction	90%	4%	6%

Source: P. Pradhan, <u>Public Works and Employment in Nepal</u>, mimeo.

ANNEX II.5 - THE INFORMAL SECTOR

Table 5.1: Estimated Informal Sector Employment – 1990 (Main Occupations)

Occupational Group	Rural % of Total	Rural Workers '000	Prop. of Informal Sector (%)	% of Total	Urban Workers '000 1990	Prop. Informal Sector (%)	Total Workers '000	Prop. of Informal Sector (%)
Economically active population > 10 yrs. ('000)	–	8261	–	–	918	–	9179	
Sales Workers	2.8%	231	25.6%	8.4	78	26.4%	308	25.8%
Service Workers	1.5%	120	13.4%	6.2	57	19.6%	178	14.9%
Production Workers	3.1%	254	28.2%	9.6	88	30.3%	343	28.7%
Transport Workers	0.1%	12	1.3%	1.9	18	6.1%	29	2.4%
General Labour	3.5%	285	31.6%	5.6	51	17.7%	336	28.2%
Total: Informal Sector	10.9%	903	100.0%	31.7	291	100.0%	1194	100.0%
Agriculture	82.2%	6791		34.9	320		7111	
Residual (formal)	6.9%	568		33.4	307		874	

NOTE: The Table presents one estimate of informal sector employment based on the information obtained from the MPHBS. The survey provides the percentage distribution of economically active persons over 62 activity groups. Breaking down the economically active population by the same proportions, and making some assumptions about which occupations fall in the informal sector, yields estimates of the number of main workers engaged in each activity. The classification of people into activity groups has been based on the concept of "main occupation". The percentages of main workers engaged in these activities have been deflated by the proportions of informal work expected to be undertaken in each activity respectively for rural and urban areas. The proportions used are based on observation and experience of rural Nepal, as well as both the plains and hill regions of neighboring states in India. The resulting distribution of main workers in the rural and urban informal sectors has then been multiplied by the numbers of economically active persons (above ten years of age).

Annex II.5

TABLE 5.2: INCIDENCE OF POVERTY AMONGST INFORMAL SECTOR WORKERS

	Poor Workers		
Occupations	Rural	Urban	Nepal
3. Sales			
3.1 Proprietors	25.6%	9.2%	18.9%
3.2 Salesmen	41.3%	7.1%	16.9%
3.3 Street vendors	20.0%	56.1%	28.4%
3.4 Supervisors/agents			
Total Sales Workers	25.4%	11.9%	19.4%
4. Services			
4.1 Hotel supervisors	31.2%	13.8%	22.2%
4.2 Hotel workers	25.5%	-	9.3%
4.3 Domestic services	7.7%	15.9%	9.2%
4.4 Caretakers/cleaners	100.0%	22.9%	27.2%
4.5 Launderers	100.0%	41.4%	57.4%
4.6 Hairdressers	28.8%	75.6%	54.2%
4.7 Security	45.8%	43.3%	44.7%
4.8 Other services		42.4%	42.4%
Total Service Workers	13.0%	22.9%	15.9%
6. Production			
6.1 Proprietors: Rice mills	24.1%	4.9%	15.3%
6.2 Textile workers	37.9%	56.9%	44.8%
6.3 Grain/spice millers	40.7%	26.6%	33.6%
6.4 Butchers	38.5%	-	21.7%
6.5 Bakers	100.0%	21.7%	35.6%
6.6 Wine/beverage makers	37.2%	40.9%	37.7%
6.7 Other food processors	-	-	-
6.8 Tobacco workers	59.3%	47.0%	53.9%
6.9 Garment workers	63.7%	30.7%	52.5%
6.10 Leathergoods workers	100.0%	-	44.1%
6.11 Furniture makers	100.0%	29.0%	49.7%
6.12 Blacksmith/tool makers	72.0%	35.6%	65.4%
6.13 Mechanics	100.0%	19.3%	26.8%
6.14 Electrical workers	-	12.2%	11.1%
6.15 Jewellery makers	56.2%	38.0%	43.7%
6.16 Printers/engravers	100.0%	19.1%	36.3%
6.17 Other prod. workers	70.9%	10.7%	42.1%
Total Production Workers	55.5%	25.6%	43.4%
7.4 Transport Workers	47.9%	8.1%	18.0%
7.5 Transport Labor	40.2%	21.2%	34.7%
Total Transport Workers	44.0%	9.8%	22.5%
7.6 General Labor	61.6%	56.2%	60.0%
Total: Informal Workers	42.0%	30.2%	37.9%
Active Population	43.5%	18.6%	37.6%

Table 5.3: Estimated Household Informal Sector Incomes by Major Sources (MPHBS, 1985)
(Rs. 1984/85 per month)

Source	Rural Rs.	%	Urban Rs.	%	Nepal Rs.	%
Monthly Total	1192	100.0	1794	100.0	1233	100.0
Income from Informal Sector						
Non-agriculture enterprise - cash and kind	64	5.4%	376	21.0%	86	7.0%
Home production	53	4.4%	69	3.8%	54	4.4%
Wages/salaries - cash and kind	45	3.8%	162	9.0%	80	6.5%
Total - Informal Sector	162	13.6%	607	33.8%	220	17.8%

Annex II.5

Table 5.4: Informal Sector Employment Projections
(thousands of workers)

Occupation Group	Workers 1990	Growth Rate (%)	Workers 2010
Optimistic Assumptions:			
Total: Informal Sector	1194	6.62%	4305
(% of total)	(13.0%)		(24.8%)
Expected Case:			
Sales workers	308	3.3%	590
Service workers	178	9.1%	1012
Production workers	343	3.4%	665
Transport workers	29	4.0%	64
General labour	336	3.6%	682
Total: Informal Sector	1194	4.7%	3013
(% of total)	(13.0%)		(17.3%)
Pessimistic Assumptions:			
Total: Informal Sector	1194	2.9%	2133
(% of total)	(13.0%)		(12.3%)

NOTE: Projections for the year 2010 of the numbers of main workers in the informal sector are presented in Table 5.4. The numbers for 1990, contained in Table 5.1, are used as the base. Three alternative sets of assumptions are used and corresponding optimistic, expected and pessimistic GDP growth rates of 4.3%, 3.3% and 2.7%. The growth rates of employment for broad categories of workers are based on a combination of historical performance in subsectors.

ANNEX II.6 - AGRICULTURE AND LAND TENURE

Annex II.6

Table 6.1: Indexes of Cultivated Area, Supply of Inputs, Production and Yield
(1975 = 100)

		Scale of Input and Area Covered				Scale of Return	
		Irrigation		HYV's			
Year	Total Cultivated Area	Annual Investment	Irrigated Area	Area Covered	Annual Sale of Fertilizer	Production	Yield Col.6/Col.1 *100
	1	2	3	4	5	6	8
1975	100.00	100.00	100.00	100.00	100.00	100.00	100.00
1976	102.24	131.66	112.78	100.58	85.62	103.44	101.17
1977	102.80	170.98	183.68	110.41	104.05	98.04	95.37
1978	103.78	193.14	211.34	153.21	124.53	94.89	91.43
1979	103.73	303.70	291.08	144.00	125.38	96.61	93.14
1980	102.80	310.16	436.72	156.35	137.97	85.18	82.86
1981	106.16	384.43	469.80	184.67	148.51	101.35	95.47
1982	108.36	479.16	520.56	218.70	157.99	105.43	97.30
1983	112.65	649.21	593.41	192.17	202.77	88.67	78.71
1984	114.80	726.12	655.93	182.07	239.05	113.53	98.89
1985	119.98	868.07	798.75	188.00	275.36	111.46	92.90
1986	124.60	1125.20	834.52	204.78	281.06	117.44	94.25
1987	124.93	1126.25	963.57	210.32	290.84	108.63	86.95
1988	134.83	1137.99	1085.57	236.58	340.05	127.16	94.31

* High yielding variety includes Paddy, Wheat and Maize.

Source: Columns 1 to 4 and 6 to 8 HMG, Nepal, 1989
Column 5 AIC, Nepal, 2045 (Poush).

Annex II.6

Table 6.2: Incidence of Irrigation by Farm Size
(Percent of Area Irrigated)

Source and Data		Large	Medium	Small	Marginal
NRB 1976/77	Terai	18.1	16.8	10.9	10.6
	Hills	7.7	7.3	3.9	7.0
MPHBS 1989	Terai	-	39.7	35.5	-
	Hills	-	36.8	27.4	-
NPC 1983/a	Terai	40.2	32.0	27.6	24.4
	Hills	21.9	25.3	23.1	20.4

/a Percentage of farm households with half or more of their land irrigated.

NOTE: Land size categories (ha.)

	NRB and NPC:		MPHBS: (Average farm size in category)	
	Terai	Hills	Terai	Hills
Large	5.1+	1.02+		
Medium	2.4-5.1	0.51-1.02	3.1	0.52
Small	1.02-2.4	0.20-0.51	1.1	0.31
Marginal	<1.02	<0.20		

Annex II.6

Table 6.3: Characteristics of Farmer Groups in the Hills

	"Large"	"Medium"	"Small" (Poor)	"Marginal" (Very Poor)
Ave. family size	8	6	5	4
Holding size (ha.)	>3.0	1.0-3.0	0.5-1.0	<0.5
% of households	5	22	19	54
Ave. holding size (ha)	5.7	1.7	0.75	0.18
Total cultivated area (%)	34	38	16	12
Household income from agriculture (%)	76	70	58	47
Family labor used on farm (%)	77	68	56	31
Total household income per month (Rs.)	1284	841	635	462
Total per capita income per month (Rs.)	162	139	128	108
Household agricultural income per month (Rs.)	973	591	372	217
Monthly agricultural income per capita (Rs.)	122	99	74	54
Share of 1985 poverty line income derived from agriculture (%)	76	61	46	34

Source: IDS (1986), <u>The Land Tenure System in Nepal</u> and MPHBS (1989).

Annex II.6

Table 6.4: Yield Assumptions for Farm Models*
(Metric Tons per Hectare)

	Terai Rainfed	Terai Irrigated Current	Terai Irrigated Foreseeable	Hill Valleys Rainfed	Hill Valleys Irrigated Current	Hill Valleys Irrigated Foreseeable	Hill Slopes Rainfed	Hill Slopes Irrigated Current	Hill Slopes Irrigated Foreseeable
Spring Paddy	-	3.00	3.80	-	3.50	4.00	-	-	-
Paddy Main	1.60	2.50	4.10	2.00	3.00	3.80	1.25	1.80	2.50
Follow Paddy	-	2.20	3.80	-	2.70	3.50	-	-	-
Upland Paddy	1.20	-	-	1.25	-	-	1.00	-	-
Maize	1.50	2.00	2.70	1.60	2.70	3.20	1.20	1.50	2.25
Wheat	0.90	2.20	2.70	1.60	2.70	3.20	1.20	1.50	2.25
Barley	0.85	1.20	1.50	0.75	1.20	1.50	1.00	-	-
Millet	1.00	-	-	1.40	-	-	0.80	-	-
Pulse	0.45	1.00	1.50	0.70	1.00	1.50	1.00	-	-
Mung	-	1.00	1.30	-	-	-	-	-	-
Oilseed	0.50	0.75	1.00	0.30	0.60	1.00	0.25	-	-
Potato	-	10.00	14.00	-	10.00	14.00	5.00	10.00	15.00

*from the Master Plan for Irrigation Development in Nepal 1989

Annex II.6

Table 6.5: Potential Income Per Hectare from Field Crops in the Hills
(1988/89 Rs. per annum)

	East Valley Floor Rs.	East Hill Slopes Rs.
Rainfed (Actual)	7,244	4,827
Current year-round irrigation (Potential)	24,569	11,903
Future year-round irrigation (Potential)	27,674	18,520

Source: Based on <u>Master Plan for Irrigation Development in Nepal</u> (1989), Cycle 1, Annexes-Volume 2.

Yield and cropping pattern assumptions are shown in Table 6.4.

The calculation of potential income from current year-round irrigation assumes that farmers have reliable access to water and chemical inputs all year. The calculation of potential income from future year-round irrigation assumes that irrigation water control improves and that farmers increase their input use accordingly.

Table 6.6: Minimum Land Holding Necessary to Earn a Poverty Line Income from Field Crops in the Hills
(hectares)

	Eastern Nepal Valley	Eastern Nepal Slope
Rainfed	1.39	2.08
Half Irrigated (current technology)	0.63	1.21
Half Irrigated (foreseeable technology)	0.58	0.86
All Irrigated (current technology)	0.41	0.85
All Irrigated (foreseeable technology)	0.36	0.54

NOTE: Based on Table 6.5, assuming a family size of 4.

Table 6.7: Estimated Household Income to be Derived from Horticultural Production on 0.18 ha.

Cropping System	Farm Income Rs.	Monthly per capita Income Rs.	Proportion of Poverty Line Income %
Banana/pineapple	5,030	105	50
Orange/lime/intercrop	8,490	177	84
Orange/lime/pear/intercrop	8,020	167	80
Apple/wheat	9,440	197	94

Source: Based on *Nepal Hill Fruit Development Project* (1987). Adapted to a farm of 0.18 ha and household size of four.

Table 6.8: Household Income and Landholdings in the Terai by Level of Poverty

	Very Poor	Poor	Non-poor
Family Size	6.33	7.14	6.76
Land Owned (ha.)	0.47	0.97	2.85
Land Operated (ha.)	0.52	1.13	2.78
Khet Land Operated (ha.)	0.24	0.95	2.30
Monthly Income Per Capita (Rs.)	79	156	340
Total Monthly Income (Rs.)	497	1113	2299
Total Monthly Agricultural Income (Rs.)	186	557	1668
Monthly Agricultural Income Per Capita (Rs.)	29	79	246
Household Income from Agriculture (%)	37%	50%	73%
Share of 1985 Poverty Line derived from Agriculture (%)	15%	40%	127%

Source: MPHBS (expressed in 1988/89 Rupees).

Annex II.6

Table 6.9: Potential Income per Hectare from Field Crops in the Terai
(1988/89 Rs./ha./yr.)

	East	West
Rainfed (Actual)	6,376	5,354
Current year-round irrigation (Potential)	18,364	16,976
Future year-round irrigation (Potential)	28,228	23,662

Source: Master Plan for Irrigation (1989).

Yield and cropping pattern assumptions shown in Tables I.4-5 and I.4-6.

The calculation of potential income from current year-round irrigation assumes that farmers have reliable access to water and chemical inputs all year. The calculation of potential income from future year-round irrigation assumes that irrigation water control improves and that farmers increase their input use accordingly.

Table 6.10: Estimates of Total Labor Use in Agriculture

Estimates of labor use per hectare are based on the Master Plan for Irrigation Development in Nepal (1989) estimates of cropping patterns and labor/hectare/crop for rainfed and irrigated land. Estimates of total irrigated land and potentially irrigated land are based on the same source.

The total number of full-time jobs in agriculture is calculated using 180 labour days/year which is somewhat lower than most estimates of full-time employment. However, the calculation of labor days/ha/year includes only labor used for individual field crops. It does not include time spent on livestock herding, fodder collection, irrigation maintenance and other activities necessary for agricultural production, but not directly related to any single crop.

All three estimates assume increased utilization of inorganic fertilizer and pesticides in areas with non-monsoon irrigation. This, in turn, requires that: (i) the input distribution system improves significantly; (ii) transportation costs in the hills do not make fertilizer use uneconomical; and (iii) farmers will be able to finance the purchase of inputs either through savings or credit.

Assumptions

Low Estimate:

Hills: No increase in areas irrigated. No new irrigation technology. Low input use in inaccessible areas. Two-thirds of irrigated land has only monsoon irrigation. Production improvements come from improved management of existing irrigation systems and increased input use in accessible areas.

Terai: No expansion in area cropped. No new irrigation technology. Production improvements come from improved management of existing irrigation systems and increased input use.

Medium Estimate:

Hills: Irrigation expanded to all valley floor land using current irrigation technology. Improved access to inputs where there is irrigation.

Terai: Some expansion in area cropped and area irrigated. No new irrigation technology.

High Estimate:

Hills: Expanded area irrigated. Improved irrigation technology. Improved access to inputs.

Terai: Expanded area cropped and irrigated. Improved irrigation technology.

Annex II.6

Table 6.10a: Low Estimate

			Cropped Hectares	Labor Days/Ha/Year	Total Labor Days/Year (000)
Hills					
Rainfed					
	Valley		316,785	210	66,524
	Slope		846,666	161	136,313
Irrigated					
	Valley				
		Accessible	40,000	408	16,320
		Inaccessible	78,000	267	20,826
	Slope				
		Accessible	23,000	383	8,809
		Inaccessible	47,000	264	12,408
		Total Hills:			261,200
Terai					
Rainfed			638,000	131	83,578
Irrigated			721,000	289	208,369
		Total Terai:			291,947
		Total Nepal:			553,147

Total number of full-time jobs in agriculture: 3.07 million

Annex II.6

Table 6.10b: Medium Estimate

		Cropped Hectares	Labor Days/Ha/Year	Total Labor Days/Year (000)
Hills				
Rainfed				
	Slope	846,666	161	136,313
Irrigated				
	Valley	273,748	408	111,689
	Slope	161,202	383	61,740
	Total Hills:			309,742
Terai				
Rainfed		400,000	131	52,400
Irrigated		1,100,000	289	317,900
	Total Terai:			370,300
	Total Nepal:			680,042

Total number of full-time jobs in agriculture: 3.79 million

Annex II.6

Table 6.10c: High Estimate

		Cropped Hectares	Labor Days/Ha/Year	Total Labor Days/Year (000)
Hills				
Rainfed				
	Slope	846,666	161	136,313
Irrigated				
	Valley	273,748	476	130,304
	Slope	161,202	428	68,994
	Total Hills:			335,611
Terai				
Rainfed		421,000	131	55,151
Irrigated		1,338,000	315	421,470
	Total Terai:			476,621
	Total Nepal:			812,232

Total number of full-time jobs in agriculture: 4.51 million

Annex II.6

Table 6.11: Estimates of Percentage of Landless Households

Region	Percent Landless	
	NPC	DSS
Mountains	3.7%	11.3%
Hills	2.2%	7.3%
Terai	18.3%	35.0%
All Nepal	10.4%	19.0%

Sources: NPC, HMG, *A Survey of Employment, Income Distribution and Consumption Pattern in Nepal*, 1983.

CBS, HMG *Demographic Sample Survey*, mimeo, 1987/88

DSS data refer to holdings of cultivated lands by rural households.

ANNEX II.7 - CREDIT AND DEBT

Annex II.7

Table 7.1: Proportion of Farm Families Borrowing from Different Sources by Farm Size

Size Group \a	Institutional 1969/70	Institutional 1976/77	Non-Institutional 1969/70	Non-Institutional 1976/77
Overall	19%	24%	82%	76%
Large	18%	34%	7%	14%
Medium	22%	21%	14%	15%
Small	64%	45%	79%	71%

a/
	Large	Medium	Small
Hills	above 1.02 ha.	0.51-1.02 ha.	Up to 0.5 ha.
Terai	above 5.42 ha.	2.71-5.42 ha.	Up to 2.71 ha.

Source: Derived from NRB (1980), Vol. I, Chapter 12, Table 2.

Annex II.7

Table 7.2: Comparative Data on Average Borrowings by Region *

Credit Agency	Hills 1976/77 Amount	Percent	Terai 1976/77 Amount	Percent
Institutional	297	33.1	532	43.6
Cooperatives and Sajhas	42	4.7	173	14.2
Agricultural Development Bank	203	22.6	271	22.2
Commercial Bank	52	5.8	88	7.2
Private	600	66.9	689	56.4
Village Moneylenders	327	36.5	245	20.1
Professional Moneylenders	19	2.1	48	3.9
Landlords	25	2.8	30	2.5
Agricultural Traders	42	4.7	77	6.3
Friends and Relatives	187	20.9	270	22.1
Others	-	-	19	1.6
Total	897	100.0	1,221	100.0

* Nepal Rastra Bank (1980), Table 12, Vol. I.

Table 7.3: Source of Loan by Ecological Region
(Percentage of Household)

Region	Bank	Cooperative	Neighbor	Moneylender	Friends	Others
Mountain	11.7	0.4	47.1	24.3	8.6	2.5
Hills	18.8	1.0	48.3	17.4	6.5	2.0
Terai	20.6	2.7	39.5	27.7	5.8	0.6
National Average	19.1	1.4	44.3	22.5	6.7	1.4

Source: CEDA, Sixth Plan Impact Evaluation Study, 1987/88.
(Note: rows do not sum to 100% in source).

Annex II.7

Table 7.4: Purpose of Borrowing for Small Farm Families
(Percent)

Purpose	1976/77 /1	1982 /2
Farming Expenditure	17.89	8.54
Capital Expenditure	36.06	23.54
Consumption Expenditure	42.37	36.32
Others	3.68	31.58
TOTAL	100.0	100.0

Note: The "controlled groups" small farmers of 1982 survey were similar to the small farm group of 1976/77 survey.

Sources: Agricultural Credit Review Survey Nepal, Derived.

1/ NRB (1980), Table 17, p. 100.
2/ NRB (1982), Table 8, p. 17.

Table 7.5: Purpose of Borrowing in Sampled Terai Village by Economic Status, 1984
(Percent)

Purposes	Low	Lower Middle	Middle	Higher	Total
Farming Expenditure	5.6	13.8	20.9	53.6	27.2
Consumption Expenditure	86.1	60.3	39.5	4.4	41.7
Social Ceremonies/ Miscellaneous Expenditures	8.3	25.9	39.5	42.0	31.1
TOTAL	100.0	100.0	100.0	100.0	100.0

Source: Yadav, (1984). Derived from Table 5.12, p. 99.

ANNEX II.8 - NUTRITION AND FOOD SECURITY

Table 8.1: Calories from Cereals and Potatoes as a Percentage of Total Calories by Income Decile by Region
(Percentage)

Per Capita Monthly Income Decile

Region	1	2	3	4	5	6	7	8	9	10
Rural Terai	88	88	89	88	86	86	84	84	83	80
Urban Terai	87	87	86	84	84	80	80	80	74	71
Rural Hills	89	89	88	89	88	86	86	85	84	84
Urban Hills	85	83	83	81	78	78	75	73	71	67
Rural Mountains	87	89	87	88	89	87	86	88	86	84

Source: MPHBS.

NOTE: Income deciles are not comparable across regions.

Annex II.8

Table 8.2: Year to Year Variation in Total Calorie Production from Grains and Potatoes (1977/78-1987/88)

Ecological Zone	Maximum Variation	Average Variation
Mountains	20%	7%
Hills	16%	9%
Terai	38%	14%
Nepal	23%	10%

Source: DFAMS.

NOTE: Variation is the absolute value of the difference in production from the previous year expressed as a percent.

Annex II.8

Table 8.3: Average Prices of Cereals by Region and Ecological Zone
(Rupees/kg: 1984/85)

REGION

Ecological Zone	Western	Central	Eastern
Mountains			
Grain/cereal	6.8	4.3	6.3
Rice Fine	8.3	5.7	7.3
Rice Coarse	8.7	5.6	7.0
Rice Other	8.9	9.1	11.2
Wheat	5.5	3.2	6.3
Maize	3.6	3.6	5.3
Coarse grains	7.7	1.3	na
Hills			
Grain/cereal	5.2	4.9	4.2
Rice Fine	6.5	6.2	5.2
Rice Coarse	6.4	5.9	4.4
Rice Other	8.2	7.5	6.2
Wheat	3.6	3.5	3.2
Maize	3.2	4.3	3.5
Coarse grains	2.5	3.5	2.8
Terai			
Grain/cereal	3.5	3.8	3.8
Rice Fine	4.3	4.4	4.2
Rice Coarse	3.7	3.9	3.9
Rice Other	4.1	5.9	5.5
Wheat	2.4	2.5	2.6
Maize	2.3	2.6	2.8
Coarse grains	na	2.2	na

Source: MPHBS.

Annex II.8

Table 8.4: Summary of District Level Food Balances (1985)
(Number of Districts)

	Surplus	Deficit (Percent of Need) 75-100%	50-75%	<50%
Region				
Mountain	3	1	5	7
Hills	5	8	16	10
Terai	14	6	0	0

NOTE: Food balance: Assumes that 80% of the NPC recommended calorie levels are provided by grains and potatoes. Population estimates for 1985 from <u>Master Plan for Irrigation Development</u> (1989). District crop production is a 4-year average of DFAMS grain and potato production statistics (1983/84- 1986/87). Milling and storage loses based on DFAMS estimates are subtracted from total production. Calorie estimates per kg of crop are based on DFAMS guidelines.

Annex II.8

Table 8.5: Prevalence of Malnutrition (percentage) in Children by Age

Age (months)	Normal	Stunted*	Wasted**
6-11	72.4	20.6	9.1
12-23	45.9	38.3	15.3
24-35	43.9	47.0	8.9
36-47	40.4	56.2	2.5
48-59	36.5	61.0	2.3
60-71	43.8	54.3	1.9
Average	45.3	51.2	6.7

Source: Nepal Nutrition Status Survey (1975).
Sample: 6500 children between 6 and 72 months old.

* Less than 90 percent of expected height-for-age.
** Less than 80 percent of expected weight-for height.

Annex II.8

Table 8.6a: Nutritional Status of Children by Total Value of Household Crop Production

	% Stunted [1]	% Wasted	N
Low (<Rs. 560)	71.6	6.3	176
Middle (Rs. 560-2050)	66.7	6.9	153
High (>R2050)	56.9	0.6	181

Children 3-10 years of age (from Martorell et al. 1984).

Table 8.6b: Nutritional Status of Children by Months of Household Food Self-Sufficiency [2]

Sufficiency of Food Production	% Stunted	% Wasted
Less than 6 months	67	5.9
6 to 9 months	54	2.1
More than 9 months	53	2.8

Total N=1490 children under 5 (SCF-Baglung (1979) cited in FAO/WFP (1987)).

Table 8.6c: Nutritional Status of Children by Area of Household Land Cultivated

Area of Land Cultivated	% Stunted [3]	% Wasted
0.0-0.5 hectares	39	9.0
0.51-1.0 hectares	36	6.5
Over 1.0 hectares	23	2.0
N	524	618

Children under 8 years of age (Nabarro, 1981)

---

[1] Stunting: less than or equal to 90 NCHS height-for-age, wasting: less than or equal to 80 NCHS weight-for-height.

[2] Definition of stunted and wasted same as above

[3] Stunted: 85% expected height-for-age, Wasted: 80% expected weight-for-age.

Table 8.7: NFC Food Grain Distribution and Transportation Costs
(1987/88)[9]

	Total Populat'n (000)	Total Foodgrain Supplied (mt)	Per Capita Foodgrain Supplied (kg)	Transport Cost (R/kg)
Region A [10] (most remote)	911	5083	5.58	8.2
Region B [11] (less remote)	1993	7601	3.81	3.0
Others [12] (accessible)	5817	11196	1.92	0.7
Kathmandu Valley	883	21700	24.58	0.4

---

[9] Adapted from IDS 'Minimum Support Price, Food Subsidy and Food Distribution Programme: Impact on the Poor'.

[10] Does not include Humla and Solokhumbu

[11] Does not include Rolpa

[12] Does not include Udayapur, Dolpa, Rasuwa, Nuwakot, Sindhupalchowk and Kabhrepalanchowk

ANNEX III

BIBLIOGRAPHY

Contents

	Page
General	215
Agriculture	219
Education	223
Population and Health	224
Women	227
Program Evaluations and Studies	228

BIBLIOGRAPHY

General

Acharya, M. <u>A Study of Rural Labor Market in Nepal</u>, Madison, Wisconsin: Dissertation Thesis, University of Winconsin, 1987.

Ackermann, Ernst. <u>Transit Cost Analysis UNCTAD (Transportation - Economic Aspects - Nepal)</u>, UNCTAD, n.d.

Adhikari, Ganesh. <u>Institutional Practices and Nepalese Poor</u>. Kathmandu: Winrock, 1987.

Agrawal, Govinda R. <u>Transport Linkages in Nepal: Prospects for Regional Cooperation</u>. Kathmandu: CEDA, 1986.

Agricultural Development Bank, Nepal. <u>Information on Subsidies</u>. n.d.

Alheritiere, Dominique. <u>Legal Aspects of Environment Policy in the Kingdom of Nepal</u>. Rome: UNDP/FAO, 1983.

Amatya, U.B. <u>A Study of Off-Farm Employment and Its Impact on Household Income and Consumption in Rural Areas of Nepal: A Case Study of Two Panchayats of Nuwakot District</u>. Kathmandu: Centre for Economic Development Administration (CEDA), 1982.

Applegate, G.B. and G.A. Gilmour. <u>Operational Experiences in Forest Management Development in the Hills of Nepal</u>. Kathmandu: International Centre for Integrated Mountain Development, (ICIMOD), Occasional Paper No. 6, 1987.

APROSC. <u>Identifying the Target Groups for Subsidizing Foodgrain Distribution in Jumla and Dailekh Districts of Nepal</u>. Draft Report, July 1989.

ARTEP Mission Report. <u>The Challenge for Nepal</u>. 1974.

Asian Regional Team for Employment Promotion (ARTEP) and International Labor Organization (ILO). <u>Employment and Basic Needs in Nepal: A Preliminary Analysis of Problems and Policies</u>. ARTEP, Bangkok, March 1982.

Bajracharya, B.B., U.B. Amatya and I. Maskey. <u>Self-Employment in the Off-Farm Sectors in Nepal</u>, Kathmandu: CEDA, n.d.

Bajracharya, Deepak. "Deforestation in the Food/Fuel Context: Historical and Political Perspectives from Nepal," <u>Mountain Research and Development</u>, 3 (3), 1983.

Bajracharya, Deepak. "Fuel, Food or Forest - Dilemmas in a Nepali Village," <u>World Development</u>, Vol. 11, No. 12, 1983.

Banskota, M. <u>Spatial Linkages and Off-Farm Employment Generation: The Case of Nuwakot District in Nepal</u>. Kathmandu: International Centre for Integrated Mountain Development (Internal Document), 1985.

Banskota, M. <u>Overview of Off-Farm Employment Generation in Nepal</u>. Kathmandu: International Centre for Integrated Mountain Development (Internal Document), 1986.

Bhandari, B., N. Kumar and B.B.S. Dongal. <u>Rural Poverty and the Poor in Nepal: Comparative Case Study of a Hill and a Terai Village</u>. Kathmandu: HMG-USAID-GTZ-IDRC-FORD-WINDROCK Project, November 1986.

Blaikie P., J. Cameron and A. Seddon. <u>The Effects of Roads in West-Central Nepal</u>. England: Overseas Development Group, University of East Anglia, 1976.

Blaikie P., J. Cameron, and A. Seddon. <u>Nepal in Crisis: Growth and Stagnation at the Periphery</u>. Bombay: Oxford University Press, 1980.

Borgstrom, Bengt-Erik. *The Patron and the Panca: Village Values and Panchayat Democracy in Nepal.* New Delhi, Vikas Publishing House Pvt. Ltd., 1980.

Campbell, G.J., R.P. Shrestha, and F. Euphrat. "Socio-economic factors in traditional forest use and management in Nepal," *Banko Jankari*, Vol 1, No. 4, pp 45-54.

Caplan, L. *Land and Social Change in East Nepal: A Study of Hindu Tribal Relations.* London: Routledge and Kegan Paul, 1970.

Caplan, P. *Priests and Cobblers: A Study of Social Change in a Hindu Village in Western Nepal.* Aylexburg (UK), International Text Book Ltd., 1972.

Chand, Diwaker. *Critical Appraisal of Rural Economy of Nepal.* Adrasha Chapakhana, Kathmandu, 1976.

Colin, S. and A. Falk. *Koshi Hills Area Rural Development Programme (KHARDEP). Nepal - A Study of the Social Economy of the Koshi Hills Area.* London: Land Resource Development Centre, 1979.

Dahal, D.R. *Rural Poverty in Nepal: Issues, Problems and Prospects.* Winrock Research Report Series, No. 6, HMG-USAID-GTZ-IDRC-FORD-Winrock Project. Kathmandu, 1985.

Department of Roads, HMG, Nepal. *Nepal Road Statistics.* 1987.

Dunsmore, J.R. *KHARDEP: Rural Development in the Hills of Nepal.* Land Resources Development Centre, Overseas Development Administration. Land Resource Study No. 36, 1987.

Environmental Resources Limited. *Natural Resource Management for Sustainable Development - A Study of Feasible Policies, Institutions and Investment Activities in Nepal with Special Emphasis on the Hills,* Parts I and II, London, 1988.

Furer-Haimendorf, C. von. *Himalayan Traders: Life in the Highland Himalaya.* London: John Murray, 1975.

Furer-Haimendorf, C. von. *The Sherpas Transformed: Social Change in a Buddhist Society of Nepal.* New Delhi, Sterling Publishers, 1984.

Gaige, Frederick. *Regionalism and National Unity in Nepal.* Berkeley, California: University of California Press, 1975.

Gilmour, D.A. *Not Seeing the Trees for the Forests: A Reappraisal of the Deforestation Crises in Two Hill Districts of Nepal.* Kathmandu: Nepal-Australia Forestry Project, 1987.

Gilmour, D.A., G.C. King and M. Hobley. *Management of Forests for Local Use in the Hills of Nepal: Changing Forest Management Paradigms.* Kathmandu: Nepal-Australia Forestry Project, 1987.

Gurung, Harka. "Distribution Pattern and Cost of Administration in Nepal," *Journal of Development and Administrative Studies,* Vol. I, No. 1, 1978.

Gurung, Harka. *Nepal - Dimensions of Development.* Kathmandu: Sahayogi Press, 1984.

Gurung, Uday and Chitanya Mishra. *Planning and Administration of Development Programmes for Disadvantaged Ethnic Groups.* Agricultural Projects Services Centre (mimeo), Kathmandu, 1982.

Hamal, Krishna B. *Rural Development Policy and Poverty in Nepal.* Rural Poverty Research Paper Series. Winrock. Kathmandu, March 1987.

His Majesty's Government of Nepal. *A Study on Bonded Labour.* Ministry for Labor and Social Welfare, Kathmandu, 1984.

His Majesty's Government of Nepal. *Rural Landlessness in Nepal.* Kathmandu, 1985.

His Majesty's Government of Nepal. Basic Needs Fulfillment in Nepal by the Year 2001. Basic Needs Task Force, Planning Commission (unpublished mimeo), 1986.

His Majesty's Government of Nepal/ILO. A Report of a Survey of Employment, Income, Expenditure and Basic Needs Satisfaction in Nepal. (Unpublished mimeo), 1986.

His Majesty's Government of Nepal. Rural Household Energy Use in the Terai and Mid-Western Hill Regions of Nepal. Vol I - Main Report. Kathmandu: Water and Energy Commission, 1987.

Hitchock, John. The Magars of Banayan Hill. New York, Holt, Rinehart and Winston, 1966.

ICIMOD. People and Jobs in the Mountains. Report of the International Workshop on Off-Farm Employment Generation in the Hindu Kush Himalayan Region (Dehra Dun, India, 17-19 May, 1986), Kathmandu.

ICIMOD. International Workshop on Off-Farm Employment Generation, Background Papers No. 3, Vol. I, Kathmandu, 1986.

ILO. Profiles of Rural Poverty, 1977.

ILO. Strategies for National Manpower Development in Nepal. Kathmandu, 1986.

Integrated Development Systems. Rural Landlessness in Nepal. April 1985.

Integrated Development Systems. An Assessment of the Government's Subsidy Policy in Nepal. Kathmandu, November 1986.

Integrated Development Systems. Land Tenure System in Nepal. 1986.

Islam, R. and R.P. Shrestha. Employment Expansion Through Cottage Industries in Nepal: Potentials and Constraints. New Delhi: ILO/ARTEP, 1986.

Islam R., A. Khan and E. Lee. Employment and Development in Nepal. Bangkok: International Labour Organization, ARTEP, 1982.

Jain, S.C. Poverty to Prosperity in Nepal. New Delhi: Development Publishers, 1981.

Joshi C., B. Pradhan and R. Prativa. Intervention for the Alleviation of Poverty Situation in Nepal: A Case of Credit Programmes. Country Paper presented in the Planning Meeting on "Women and Poverty: State Policies, Grass-Root Initiatives and Responses" Asia and Pacific Development Centre, Kuala Lumpur, Malaysia and Women Development Section, Kathmandu, October 3-7.

Joshi C., B. Pradhan and R. Prativa. Employment Expansion through Cottage Industries in Nepal: Potentials and Constraints. ILO-ARTEP, New Delhi, 1986.

Kayastha, N. Employment in Hotel Industry of Nepal, Kathmandu: CEDA, 1985.

Koirala, Bachchu Prasad. Economics of Land Reform in Nepal: A Case Study of Dhanusha District. Kathmandu: HMG-USAID-GTZ-IDRC-FORD-WINROCK Project, 1987.

McDougal, Charles. Village and Household Economy in Far Western Nepal. Kathmandu, Tribhuvan University, 1968.

McFarlane, Alan. Resources and Population: A Study of the Gurungs of Nepal. London: Cambridge University Press, 1976.

McNeely, J. Man and Nature in the Himalaya. What can be done to ensure that both can prosper? Paper presented at the International Workshop on the Management of National Parks and Protected Areas of the Hindu-Kush Himalaya, Kathmandu, 1985.

Ministry of Health. _Development Budget, Fiscal Year 2043/44_. 1987.

Ministry of Finance. _Economic Survey, Fiscal Year 1985/86_. 1987.

Molnar, A. _The Dynamics of Traditional Systems of Forest Management_. Washington, D.C.: Development and Training Project, World Bank, 1981.

National Planning Commission. _The Sixth Plan (1980-85)_. Kathmandu, 1981.

National Planning Commission. _The Seventh Plan (1985-90)_. Kathmandu, 1981.

National Planning Commission. _A Survey of Employment, Income Distribution and Consumption Patterns in Nepal_. Kathmandu, 1983.

National Planning Commission. _Program for Fulfilment of Basic Needs (1985-2000)_. Kathmandu, 1987.

National Planning Commission Secretariat. _Census of Manufacturing Establishments: Nepal 1986-87_. Kathmandu: Central Bureau of Statistics, 1988.

Nepal Institute of Development Studies (NIDS). _Wage Structure in Nepal_. A Survey Report, Kathmandu, Nepal, 1984.

Nepal Rastra Bank. _Multipurpose Household Budget Survey - A Study on Income Distribution, Employment and Consumption Pattern in Nepal_, Vol. I, 1985.

Nepal Land Resources Mapping Project. _Economics Report for the Far Western Development Region_. Kathmandu, December 1981.

New Era/UNICEF. _Children of Nepal: A Situation Analysis_. Kathmandu, 1981.

New Era. _Food Systems and Society in Nepal: An Overview_. Kathmandu, 1981.

New Era. _Development of Small Towns in Gandaki Growth Axis, Central Nepal_. Kathmandu, 1986.

New Era. _A Survey of Informal Sector in Kathmandu_, (Unpublished), Kathmandu, 1986.

New Era. _Education, Manpower and Employment_, Kathmandu, 1988.

Panday, D.R. "Foreign Aid in Nepal's Development" in Integrated Development Systems, ed., _Foreign Aid and Development in Nepal_. Kathmandu: IDS, 1983.

Paneru, Sudha. _Traditional and Prevailing Child-Rearing Practices Among Different Communities in Nepal_. Kathmandu, Center for Nepal and Asian Studies, 1980.

Poffenberger, Mark. _Patterns of Change in the Nepal Himalaya_. New Delhi: The Macmillan Company of India Limited, 1980.

Pradhan, Bina, Indira Shrestha and Ananda Shrestha. _Child Development Study in Nepal: Implications for Policy and Training_. Center for Economic Development and Administration, Kathmandu, 1980.

Raj, Kaplana and Laxmi Sayenja. _A Micro Perspective on Poverty: A Case Study of Kubinde Village Panchayat, Sindhupalchowk District_. Kathmandu: HMG/WINROCK Project, 1987.

Regmi, Mahesh C. _Land Ownership in Nepal_. Berkeley, California: University of California Press, 1985.

Schaffner, Urs. _Road Construction in the Nepal Himalaya: The Experience from Lamo-Samgu - Jiri Road Project_. Kathmandu: ICIMOD, 1987.

Seddon, D., P. Blaikie and J. Cameron. Peasants and Workers in Nepal. Warminister, Wilts, England: Aris and Phillips Ltd., 1979.

Seddon, David. Nepal: A State of Poverty, New Delhi: Vikas Publishing House, 1987.

Seth, A.N. Report of Land Reforms in Nepal, Bangkok: FAO, 1967.

Sharma, K. Chandra. Natural Resources of Nepal. Calcutta, India: Navana Printing Works Pr. Ltd., 1978.

Shrestha, Evaluation of Land Reform Programme in Nepal, Kathmandu: Tribhuvan University, 1978.

Shrestha, R.R. "Small-Scale Industries and Employment Generation in Nepal." Kathmandu: ICIMOD (Internal Document), 1985.

Singh, R.B. A Review of Nepal's Efforts in Poverty Alleviation. Rome: FAO, February 1983.

UNDP and World Bank. Energy Assessment Status Report. Kathmandu: UNDP, 1985.

UNICEF/HMG. Plan of Operations: Basic Services for Children and Women in Nepal, 1988-1992.

World Bank. Nepal - Policies for Improving Growth and Alleviating Poverty. 1988.

World Bank, IDA Mission. Nepal - Social Sector Strategy Review, Draft Report, 1989.

Zaman, M.A. Evaluation of Land Reform in Nepal. Kathmandu: HMG Press, 1982.

Agriculture

Acharya, Shiva Prasad. Country Paper on Integrated Rural Development in Nepal, Kathmandu: APROSC, CIRDAP, March 1982.

Acharya, Meena. "Surplus Labor in Nepalese Agriculture," The Journal of Development and Administrative Studies, Vol. I, No. 2 (January), Kathmandu, pp. 222-243. 1979.

Agricultural Input Corporation, Seed Production and Input Storage Project. To Serve Hill Farmers with Improved Seeds and Farming Inputs. Kathmandu, 1985.

Agricultural Development Bank/Nepal. A Case Study of Impact of Tractor and Pumpset Loans (Outside Project Areas). Kathmandu, Nepal. 1973.

Agricultural Development Bank/Nepal. Impact of Tractor and Pumpset Loan on Resource Allocation, Farm Production Profitability in Bhairahawa Project, Kathmandu, November 1973.

Agricultural Development Bank/Nepal, Planning Research and Project Division. Impact of Mechanization and Irrigation and its Policy Implication. Kathmandu, January 1983.

Agricultural Development Bank/Nepal. A Study of the Impact of Water Turbines in Nepal. Agricultural Development Bank/Nepal, Evaluation Division, Head Office, Kathmandu, 1987.

Alirol, P. Transhuming Animal Husbandry Systems in Halingchowk Region (Central Nepal):. A Comprehensive Study of Animal Husbandry on the Southern Slope of the Himalayas. Bern: Swiss Development Cooperation, 1979.

APROSC. Agrarian Reform and Rural Development in Nepal, Country Review Paper, Kathmandu, February 1978.

APROSC. A Report of the National Level Food Policy Seminar, 1984.

APROSC. A Preliminary Study on Private Sector Participation in Cereal Seed Industry in Nepal. Kathmandu, 1985.

APROSC. Nepal - State-of-the-Art Report on Integrated Rural Development, (Jagadish R. Baral/Kiran Koirala), March 1987.

APROSC. Subsistence Farm Income and Resource Allocation. Kathmandu, 1988.

Asian Development Bank. Nepal Agriculture Sector Strategy Study. Detailed Sector Review and Analysis, Vol. II, 1982.

Asian Development Bank. Co-operation with NGO's in Agriculture and Rural Development in Nepal, March 1989.

Axinn, Nancy and George Axinn. Small Farmers in Nepal: A Farming Systems Approach to Development, Kathmandu, n.d.

Banskota, Mahesh. Hill Agriculture and the Wider Market Economy. Kathmandu: ICIMOD Occasional Paper No. 10, May 1989.

Bhadra, Binayak. Analysis of Complementarities amongst Irrigation and Fertilizer Projects Investment. Kathmandu, CEDA, 1982.

Bista, Ramesh. Cross-Sectional Variations and Temporal Changes in Land Area Under Tenancy and their Implications for Agricultural Productivity. Kathmandu: WINROCK Project, 1989.

Cassels, C., A. Wijga, M. Pant, D. Nabarro. Coping Strategies of East Nepal Farmers: Can Development Initiatives Help? KHARDEP Impact Studies, Supplementary Report, October 1987.

Central Bureau of Statistics. A Report of an Analytical Study of the National Sample Census of Agriculture of Nepal,1981/82, Kathmandu, 1982.

Central Bureau of Statistics. A Comparative Study of the National Sample Census of Agriculture of Nepal, Kathmandu, 1986.

CIWEC. Master Plan for Irrigation in Nepal, 1989.

Dahal, Dilli R. "Population Growth, Land Pressure and Development of Cash Crops in a Nepali Village" in Contributions to Nepalese Studies. Vol 13, No. 1.

Dani, A.A. and G.J. Campbell. Sustaining Upland Resources - People's Participation and Watershed Management. International Centre for Integrated Mountain Development (ICIMOD), Occasional Paper No.3, 1986.

Department of Food and Agricultural Marketing Services. Farm Management Study in the Selected Districts of the Hills and Terai of the Central Development Region of Nepal, 1981/82. Kathmandu, 1981/82.

Department of Food and Agricultural Marketing Services. A Study on Impact of Subsidized Food Grains Distribution on the Production of Main Agricultural Products in Nepal, (IDS) Kathmandu, 1984.

Department of Food and Agricultural Marketing Services. Balance Sheet: Food Grain for the Kingdom of Nepal, 1984/85, 1985.

Department of Food and Agricultural Marketing Services. The Cost of Production of Major Crops in Nepal. Ministry of Agriculture, Kathmandu, 1986.

Department of Food and Agricultural Marketing Services. Main Report on National Farm Management Study, Nepal, 1983-85. Ministry of Agriculture, Economic Analysis Division, 1986.

FAO/FADINAP. Proceedings of National Seminar on Pricing Policies and Transport Studies, 1988.

FAO/WFP. *Assessment of Food and Agricultural Situation*, 1987.

Feldman, David. *Compost or Chemicals: Technical Choice in the Hill Agriculture of Nepal*. Overseas Development Group, Norwich. July 1974.

Galt, D.L. and S.B. Mathema. *Farmers' Participation in Farming System Research* (SERED Report No. 3), Department of Agriculture, October 1986.

Gianner, Hans J. *Farming Systems Research Tansen: Tinau Watershed Project*, HMG/SATA, 1987.

Hamal, K.B. *Risk Aversion, Risk Perception and Credit Use: The Case of Small Paddy Farmers in Nepal*. APROSC/ADB, 1983.

Holman, L. Dale. *Observations and Recommendations on Women's Agricultural Extension Programs in Hill Districts of Myagdi, Baglung and Parbat*, Winrock International Institute for Agricultural Development/Nepal, Submitted to ARPP, December 1986.

Hopkins, Nigel. *The Fodder Situation in the Hills of Eastern Nepal*. APROSC Occasional Paper 2. 1983.

Jamison, D.T. and P.R. Moock. *Farmer Education and Farm Efficiency in Nepal, The Role of Schooling, Extension Services and Cognitive Skills*, Washington, D.C. The World Bank, 1984.

Katwal, R.B. *Wages and Welfare: The Case of Attached vs. Casual Labour in the Nepal Tarai*, Research Paper Series, No. 31: HMG-USAID-WINROCK Project, Kathmandu, Nepal.

Khadga, B.B. and J.C. Gautam. "Demand and Production of Food Grains in the Hills" in Ministry of Agriculture, *Nepal's Experience in Hill Agricultural Development*. Kathmandu, 1981.

Khadka, K.R. and I.R. Lohani. *Effects of Population Pressure on Land Tenure, Agricultural Productivity and Employment in Chitwan and Tanahu Districts*, CEDA, Kathmandu, 1982.

Khatwal, Bhimendra B. *Wages and Welfare: The Case of Attached vs. Casual Labor in the Nepal Tarai*," Research Paper Series No. 81. HMG-USAID-WINROCK Project. Kathmandu, Nepal, 1986.

Leslie, J. *The Adoption of Improved Maize Varieties in Pakhribas Local Target Area: Results of a Survey and their Implications for further research*, Pakhribas Agricultural Centre, Tech. Paper No. 88, 1986.

Martin, E.D. *Resource Mobilisation, Water Allocation, and Farmer Organisation in Hill Farmer Irrigation Systems in Nepal*. Ithaca, N.Y.: Ph.D. Thesis, Cornell University, 1986 (Unpublished).

Mudbhary, P.K. *Food Supply and Distribution System in Nepal*. APROSC. 1981.

Mughes, D. *Experiences with multi-disciplinary extension at PAC and PAC's link with KHARDEP using subject matter specialists*, Pakhribas Agricultural Centre, June 1982.

Nepal Rastra Bank. *Agricultural Credit Review Survey*. Kathmandu, 1980.

Nepal Rastra Bank. *A Study on Institutional Loan Delinquency in Agriculture Sector*. Kathmandu, 1985.

Pachico, D.H. *Small Farmer Decision Making: An Economic Analysis of Three Farming Systems in the Hills*. Ithaca, N.Y.: Cornell University, Ph.D. Thesis (unpublished), 1980.

Panday, K.K. *Fodder Trees and Tree Fodder in Nepal*. Switzerland: Swiss Development Cooperation and Federal Institute of Forestry Research, 1982.

Panday, K.K. *The Livestock Fodder Situation and the Potential for Additional Fodder Resources*, Mountain Environment and Development, SATA, 1982.

Pant, Mahesh, Alet Wijga, Claudia Cassels, and David Nabarro. Report of the Kosi Hill Area Development Programme (KHARDEP), Impact Studies: Final Report on the Use of Agricultural Inputs and Extension Services in the Kosi Hill Area, East Nepal. KIS Supplementary Report No. 4. Department of International Community Health. Liverpool, July 1986.

Pokharel, B.N. and G.P. Shivakoti. Impact of Development Efforts on Agricultural Wage Labour. WINROCK Project. Kathmandu, 1986.

Pradhan, M.L. and B.P. Sinha. Agricultural Extension in Nepal, paper presented at ICIMOD Workshop on Mountain Development, January 1989.

Pradhan, Bharat B. Medium Irrigation Project: Participatory Irrigation. IBRD, Kathmandu, October 1982.

Pratap, Narayan. Nepal Fertilizer Pricing and Subsidy Policies, 1987.

Pudasaini, S. P. Farm Mechanisation, Employment and Income in Nepal: Traditional and Mechanised Farming in Bara District, Kathmandu: HMG-A/D/C Research Paper Series, No. 3. HMG-A/D/C Project, 1980.

Pudasaini, S.P. Production and Price Responsiveness of Crops in Nepal. ADC/APROSC, Kathmandu, 1984.

Rauniyar, Krishna K. Labor Utilization and Non-Farm Labor Supply Among Rural Farm Households: A Case Study of Hill and Tarai Districts, Research Paper Series, No. 30. Kathmandu: HMG-USAID-GTZ-WINROCK Project, 1985.

Rokaya, Chandra M. Impact of the Small Farmers Credit Programme on Farm Output, Net Income, and the Adoption of New Methods: A Nepalese Case Study. (edited by Som P. Pudasaini), Armidale, University of England, January 1983.

Schroeder, R.F. "Himalayan Subsistence Systems: Indigenous Agriculture in Rural Nepal" in Mountain Research and Development, Vol. 5, No. 1, pp. 31-44, 1985.

Sharma, Manandhar. Factors Affecting the Tarai Paddy Market: Pricing Policy Implications, Winrock, 1986.

Sharma, R.P. and J.R. Anderson. Nepal and the CGIAR Centres. CGIAR Study Paper Number 7. World Bank, 1985.

Sharma, S. Agrarian Change and Agricultural Labour Relations, Kathmandu: Winrock, 1987.

Sharma, Shiv P. Share Cropping and Input Sharing in Nepal. APROSC mimeo, Kathmandu, 1982.

Sharma, Shiv P. Land Owners' Output Share and Factor Payments to Land: A Case Study for Nepalese Share Croppers. APROSC mimeo, Kathmandu, 1983.

Sharma, Shiv P. Tenancy Issues in Nepal. APROSC mimeo, Kathmandu, 1983.

Sharma, Shiv P. and Jawahar Lal Amatya. Irrigation and Land Tenure in Nepal: A Quick Survey. APROSC. September 1983.

Sharma, Shiv P. Rural Real Wage Rate in Nepal: A Time Series Analysis, HMG-USAID-GTZ-IDRC-FORD-WINROCK Project, 1987.

Shrestha, P., L. Ziveta, B. Sharma and S. Anderson. Planning Extension for Farm Women. Final Report, Integrated Cereals Project, Department of Agriculture. Kathmandu, 1984.

Shrestha, S.P. Community-Managed Irrigation Systems: Case Study of Arughat-Vishal Nagar Pipe Irrigation Project, Winrock, 1987.

Shrestha, R.L.J. and D.B. Evans. "The Impact of Forestry Project on Livestock Production: Some Evidence from Nepal" in <u>Journal of Agricultural Economics</u>, Vol. XXXV, No. 2, May 1987, pp. 273-278.

Solid Waste Management Project. <u>Communication Techniques and Evaluation of Composting Management</u>, August 28, 1987, (SWMP).

Thapa, G.B. <u>The Impact of New Agricultural Technology on Income Distribution in the Nepalese Tarai</u>. Ithaca, N.Y.: Ph.D. Dissertation, Cornell University, 1989.

Thapa, H., T. Green and D. Gibbon. <u>Agricultural Extension in the Hills of Nepal: Ten Years of Experience from Pakhribas Agricultural Centre</u>. London: Overseas Development Institute, Agricultural Administration Unit, 1988.

Wallace, M. <u>Fertilizer Price Policy in Nepal</u>. IRDC/APROSC. Kathmandu, 1985.

Wallace, M.B. <u>Food Price Policy in Nepal</u>. Winrock, Kathmandu, 1987.

WECS. <u>Irrigation Sector Review</u>. Water and Energy Commission Secretariat, Kathmandu, 1982.

Yadav, R.P. <u>The Concept of Integrated Rural Development</u>. Kathmandu, APROSC, CIRDAP, March 1982.

Yadav, R.P. "On Farm and Off Farm Employment Linkages," Kathmandu: ICIMOD (Internal Document), 1986.

Yadav, R.P. <u>Agricultural Research in Nepal: Resource Allocation - Structure and Incentive</u>, IFPRI Research Report, Washington, D.C., 1987.

Zevering, K.M. <u>Agricultural Development and Agrarian Structure in Nepal</u>. Bangkok: ILO/ARTEP, 1974.

Education

Asian Development Bank and His Majesty's Government. <u>Nepal Education Sector Study</u>. Asian Development Bank, 1982.

Bennett, Nicholas. "Empty Benches in Primary Schools" in <u>Development International</u>, January/February 1987.

CERID (Research Centre for Educational Innovation and Development). <u>Achievement Study of Primary School Children</u>. Kathmandu: Tribhuvan University, 1980.

CERID. <u>Meeting Educational Needs of Young People Without Schooling or with Incomplete Schooling</u>. Kathmandu: Tribhuvan University, 1982.

CERID. <u>Parents' Attitudes Towards and Expectations from Education: A Seminar Report</u>. Kathmandu: Tribhuvan University, 1982.

CERID. <u>Primary Education in Nepal: Progress Towards Universalization</u>. Kathmandu: Tribhuvan University, 1983.

CERID. <u>Determinants of Educational Participation in Nepal</u>. Kathmandu: Tribhuvan University, 1984.

CERID. <u>Education and Development</u>. Kathmandu: Tribhuvan University, 1984.

CERID. <u>Education of Girls and Women in Nepal</u>. UNICEF, Kathmandu, 1984.

CERID. <u>Effectiveness of Primary Education in Nepal</u>. National Planning Commission, Kathmandu, 1986.

CERID. <u>Women's Participation in Non-formal Education Programme in Nepal</u>. Kathmandu: Tribhuvan University, 1986.

CERID. <u>An Inquiry into the Causes of Primary School Drop-out in Nepal.</u> Kathmandu: Tribhuvan University, 1987.

CERID. <u>Primary Education Project - An Evaluation Study Report.</u> Kathmandu: Tribhuvan University, 1987.

CERID/IDRC. <u>Instructional Improvement in Primary Schools.</u> Kathmandu: Tribhuvan University, 1986.

CERID/Ministry of Education and Culture. <u>Promotions of Girls' Education in the Context of Universalization of Primary Education.</u> Kathmandu: Tribhuvan University, 1985.

Ministry of Education and Culture/USAID. <u>Improving the Efficiency of Educational Systems - Education and Human Resources Sector Assessment: Nepal.</u> Coordinated for HMG/N by the MOEC with USAID, Florida State University, Tallahassee, Florida, U.S.A., 1988.

Ministry of Education and Culture. <u>Education for Girls and Women in Nepal: An Overview.</u> Kathmandu, 1986.

Ministry of Education and Culture, Manpower and Statistics Section. <u>Education Statistics Report of Nepal, 1987.</u>

Shrestha, Gajendra Man, et al. "Determinants of Educational Participation in Rural Nepal" in <u>Comparative Education Review</u>, November 1986.

Shrestha, Kedar. <u>Educational Experiments in Nepal.</u> Kathmandu: Tribhuvan University, 1982.

UNICEF. <u>Impact of the Free Textbook Distribution Programme on Primary School Enrolment in Nepal.</u> Kathmandu, 1982.

Wheeler, C. <u>The Role of Supervision in the Teaching-Learning Process in Nepal.</u> Institut International pour l'Education Primaire. Paris, 1980.

WSCC. <u>An Inventory and Analysis of Women Development Projects: Part I.</u> Kathmandu, 1983.

Population and Health

Bennett, Lynn. <u>Pregnancy, Birth and Early Child Rearing: Health and Family Planning - Attitudes in a Brahman-Chhetri Community</u>, Paper No. 9, Department of Local Development/UNICEF. Kathmandu, 1974.

Cardinalli, Robert. <u>Land, Migration and Institutional Adaptation in the Jumla Region of Western Nepal.</u> Paper presented at the International Symposium on Mountain Population Pressure, Kathmandu, 1983.

Central Bureau of Statistics. <u>Demographic Sample Survey of Nepal, 1974/75.</u> Kathmandu, 1978.

Central Bureau of Statistics. <u>Total Mortality and Fertility Rates in Nepal.</u> Kathmandu, 1978.

Central Bureau of Statistics. <u>Population Census, 1981.</u> Kathmandu, 1984.

Central Bureau of Statistics. <u>Intercensal Changes of Some Key Census Variables, Nepal 1952/54-81.</u> Kathmandu, 1985.

Central Bureau of Statistics. <u>Population Census, 1981.</u> Kathmandu, 1985.

Central Bureau of Statistics. <u>Population Projection of Nepal (Total and Sectoral), 1981-2001.</u> Kathmandu, 1986.

Central Bureau of Statistics. <u>Demographic Sample Survey of Nepal, 1986/87.</u> Kathmandu, 1987.

Central Bureau of Statistics. <u>Population Monograph of Nepal.</u> Kathmandu, 1987.

Dahal, Dilli Ram, Navin K. Rai and Andrew E. Manzardo. <u>Land and Migration in Far Western Nepal.</u> Kathmandu, Centre for Nepal and Asian Studies, Tribhuvan University, 1977.

Daly, Patricia. Health Financing and Cost Recovery in Nepal - Integrated Rural Health and Family Planning Services, Kathmandu: Project No. 367-0135, JSI, 1987.

Dhakal, Ramji, Susie Graham-Jones and Geraldine Lockett. Traditional Healers and Primary Health Care in Nepal. Kathmandu: Save the Children Fund (UK), 1986.

Family Planning and Maternal and Child Health Programme (FP/MCHP). Nepal Contraceptive Prevalence Survey Report, 1981, Kathmandu, 1982.

FP/MCH Project. Nepal Fertility and Family Planning Survey. Kathmandu, 1986.

FP/MCH Project. FP/MCH Acceptors Profile Statistics, 1983/84. Kathmandu, n.d.

FP/MCH Project. Nepal FP/MCH Data Analysis. Final Report. The Population Council, n.d.

Gubhaju, Bhakta B. "Regional and Socio-economic Differentials in Infant and Child Mortality in Rural Nepal," in Contributions to Nepalese Studies, Vol XIII, No. 1. Kathmandu, 1985.

Gubhaju, B. B. "Population Development and Family Planning," in Integrated Development Systems, ed. Status Report on Population and Development in Nepal, Kathmandu, 1986.

Integrated Development Systems. The Symbiotic Relationship Between Population, Forestry and Energy: The Case of Far-Western Development Region. Kathmandu, 1981.

Integrated Development Systems. Health and Utilization of Health Services Facilities Part II: Supplementary Data Analysis. Report submitted to USAID/NEPAL, Kathmandu, 1985.

JICA. Basic Survey Report on Population and Family Planning in the Kingdom of Nepal. Kathmandu, 1986.

Joshi, P.L. and A.S. David. Demographic Targets and Their Attainments: The Case of Nepal. Kathmandu, National Planning Commission, 1983.

Justice, Judith. Policies, Plans and People - Culture and Health Development in Nepal. Berkeley, Calfornia: University of California Press, 1986.

Karki, Y.B. Estimates of TFM (Trends in Fertility and Mortality) for Nepal and its Geographical Sub-divisions and Administrative Zones. Kathmandu: National Planning Commission, 1985.

Kochupillai, N. and C.S. Pandav. Iodine Deficiency Disorders: Nepal. New Delhi, Aide Medicale et Sanitaire/UNICEF, 1985.

Malla, Dibya. Study on Causes of Maternal Death in Nepal. Kathmandu, WHO, 1986.

Manandhar, M. "Trends, Problems and Policies for Population and Employment" in S. Manandhar, and S.J. Rana, Employment and Population Planning, Kathmandu: CEDA, 1979.

Martorell, Reynaldo, J. Leslie and P. Moock. "Characteristics and Determinants of Child Nutritional Status in Nepal," The American Journal of Clinical Nutrition, Vol. 39, 1984.

MOH/HMG and WHO. Country Health Profile Nepal. Policy Planning, Monitoring and Supervision Division, Ministry of Health, Nepal, 1988.

Nag, M., B.N.F. White and R.C. Peet. "An Anthropological Approach to the Study of the Economic Value of Children in Java and Nepal," Current Anthropology, Vol. 19, No. 2, 1978.

National Planning Commission. Internal and International Migration in Nepal. 6 parts. Kathmandu, 1983.

National Planning Commission. Review of the Delivery of Health Services Including Family Planning and Maternal and Child Health, Kathmandu, 1985.

National Nutrition Coordination Committee (NNCC). National Nutrition Strategies. Kathmandu, 1978.

New ERA. A Study on Population Planning and Maternal and Child Health, UNFPA. Kathmandu, 1986.

New ERA. Fertility and Mortality Rates in Nepal. Kathmandu, 1986.

NMEO. Report of an Analysis of the Nepal Malaria Eradication Organization Activities, 1984-86, 1988.

Pant, Y.P. Population Growth and Employment Opportunities in Nepal. New Delhi, Oxford and IBH Publishing Co., 1983.

Rajbhandari, A.K. Population Education through Women Entrepreneurship. Kathmandu: CIDB, Cottage Industry for Women Project, n.d.

Ravindran, Sundari. Health Implications of Sex Discrimination in Childhood: A Review Paper and Annotated Bibliography. Geneva, WHO/UNICEF, 1986.

Shrestha, Vijaya. Family Level Perceptions of Nutrition, MCH and Primary Health Care Programmes in Sindhu-Palchowk, Nepal. Paper presented at the Workshop on Effects of National Programmes of Nutrition and Primary Health Care on the Health-Seeking Behaviour of Families, United Nations University, Bellagio (Italy), July 1985.

Singh, M.L. and K. Pradhan. Report on the Survey of Disabled Persons in Nepal. Kathmandu: Handicapped Services Co-ordination Committee, 1980.

Stone, Linda. "Concepts of Illness and Curing in a Central Nepal Village," Contributions to Nepalese Studies, Vol. III, Kathmandu, 1976.

Thapa, S. and R.D. Rutherford. "Infant Mortality Estimates Based on the 1976 Nepal Fertility Survey" in Population Studies, Vo. 36, No. 1, 1982.

Tuladhar, J.M. "Determinants of Contraceptive Use in Nepal," Journal of Biosocial Science, Vol 17, 1985.

Tuladhar, J.M., B.B. Gubhaju and J. Stoeckel. Population and Family Planning in Nepal, Kathmandu: Ratna Pustak Bhandar, 1978.

U.S. Department of Health, Education and Welfare. Public Health Services Centre for Disease Control, and USAID, Nepal Nutrition Status Survey. Kathmandu, 1975.

UNFPA. Report on the Evaluation of UNFPA-assisted Women, Population and Development Projects in Nepal. Kathmandu, 1985.

UNICEF. Inter-Regional Migration in Nepal. Kathmandu, 1981.

World Bank. Kingdom of Nepal: Report on Population Strategy. Kathmandu, 1985.

Wright, Nicholas. Epidemiological Review of Data on Primary Health Problems in Nepal. Kathmandu: John Snow Public Health Group Inc., 1986.

Women

Acharya Meena and Lynn Bennett. "Women and the Subsistence Sector: Economic Participation and Decision Making in Nepal," World Bank Working Paper Series 526, Washington, D.C., 1983.

Acharya, Meena and Lynn Bennett. The Rural Women of Nepal: An Aggregate Analysis and Summary of 8 Village Studies. Vol. II, Part I, in M. Acharya and L. Bennett, eds. The Status of Women in Nepal. Kathmandu: Center for Economic Development and Administration, 1981.

Basnet, S. Farming, Carpet Weaving and Women: A Case Study of Bishnudevi Village, Panga, Dissertation: Tribhuvan University, Nepal, n.d.

Bennett, Lynn. Dangerous Wives and Sacred Sisters: The Social and Symbolic Roles of Women among Brahmans and Chhetris of Nepal. New York, Columbia University Press, 1981.

Cameron, M. "Landholding, Caste, and Women's Autonomy: A Study of Low-Caste Families in Rural Nepal," Dissertation Research Interim Report: Michigan State University, 1987.

Gurung, M. An Analytical Study of Time Allocation of Married Female Workers of Some Selected Industries in Kathmandu: Dissertation: Tribhuvan University, Nepal, 1981.

IFAD. Rural Women in Agricultural Investment Projects, 1977-1984. Prepared for the Nairobi Conference on the Evaluation of U.N. Decade for Women, 1985.

Integrated Development Systems. A Study of Female Employment and Fertility in One District of Nepal, Kathmandu, 1982.

Jones, S.K. Domestic Organization and the Importance of Female Labour Among the Limbu of Eastern Nepal. Ph.D. Thesis: State University of New York at Stony Brook, 1977.

Joshi, A. Women's Participation in Carpet Industry: With Special Reference to Kathmandu District, Dissertation: Tribhuvan University, Nepal, 1985.

Joshi, C. "Women in Nepal and Off-Farm Employment," Kathmandu: ICIMOD (Internal Document), 1985.

Manandhar, B. Fertility History of Working Women, Dissertation: Tribhuvan University, Nepal, 1985.

Ojha, H.K. Women's Participation in Hand Loom Industry at Kirtipur Village Dissertation: Tribhuvan University, Nepal, 1984.

Pradhan, Bina. "Foreign Aid and Women" in Integrated Development Systems, eds. Foreign Aid and Development in Nepal, Kathmandu, 1983.

Rana, M.S.J.B. and G.M. Rana. Role of Women in Nepal's Industrial Development: Status, Constraints, Opportunities and Prospectus: HMG/Nepal-UNIDO/Vienna Project, 1987.

Shaha, Iswari. Loans to Women under the Small Farmers Development Programme, Kathmandu: HMG-USAID-GTZ, Winrock Project, 1985.

Shrestha, Neeru. Women's Employment in Industrial Sector - Nepal. Kathmandu, CEDA, 1983.

Shrestha, Neeru. An Analysis of Women's Employment in Financial Institutions - Nepal. Kathmandu: CEDA, 1982.

Thapa, K.B. Women and Social Change in Nepal, Kathmandu, 1985.

UNICEF. Children and Women of Nepal: A Situation Analysis, Kathmandu, 1987.

UNIDO. The Current and Prospective Contribution of Women to Nepal's Industrial Development. Regional and Country Studies Branch, Industrial Policy and Perspective Division, 1988.

Program Evaluations and Studies

Agricultural Development Bank of Nepal. A Decade of Small Farmers Development Program in Nepal, Kathmandu, 1986.

Agricultural Development Bank of Nepal. A Study on Cost of Borrowing (ADB/Nepal Loans), Monitoring and Evaluation Division, October 1986.

Agricultural Development Bank of Nepal. Case Studies on Small Farmers Development Programme, Nepal. 1988.

Agricultural Development Bank of Nepal. Preliminary Proposal for Financing Small Farmers Development Programme, Nepal. 1988.

APROSC. Evaluation Study of Agriculture Inputs and Credit Services in Nepal. Kathmandu, 1978.

APROSC. A Report of the Rapti Baseline Survey. 1980.

APROSC. Socio-economic Benchmark Survey of Babai Irrigation Project (Bardia District). Kathmandu, October 1982.

APROSC. Crop Intensification Programme (Bara, Parsa and Rautahat), Household Base-line Study, 1983.

APROSC. Socio-economic Study of Second Command Area Development Project (Chandra Canal, Marchawar and Mohana). June 1983.

APROSC. Socio-economic Benchmark Study of Mahakali Irrigation Project (Stage I), Kathamandu, September 1983.

APROSC. Water Management in Nepal. Seminar on Water Management Issues held in Kathmandu, July 31-August 2, 1983. Sponsored by Ministry of Agriculture. Agricultural Project Services Centre, and Agricultural Development Council, Kathmandu.

APROSC. Nepal: The Hill Food Production Project (Gorkha, Lamjung, Tanahun and Syangja), Household Base-line Study. December 1984.

APROSC. Household Base-line Survey, Agricultural Extension and Research Project (Dhanusa, Mahottari and Sarlahi District). September 1985.

APROSC. Household Base-line Survey, Agricultural Extension and Research Project (Bardia, Kailali and Kanchanpur Districts). June 1986.

APROSC. Socio-economic Benchmark Study of Sunsari-Morang Irrigation Project (Stage II). Kathmandu, 1986.

APROSC. Evaluation of IFAD-assisted Small Farmer Development Programme. 1987.

APROSC. Monitoring Survey Report of Bhairahawa-Lumbini Ground Water Project Second Report. Kathmandu, 1987.

APROSC. Socio-economic Impact Study of Sunsari-Morang Irrigation Project (Stage I), Kathmandu, 1987.

APROSC. Impact Evaluation of the Asian Development Bank Assistance at Farm Level in Nepal: A Case Study on Shallow Tubewells. June 1988.

APROSC. Impact Evaluation of Bhairahawa-Lumbini Ground Water Project, Stage I, Final Report. Kathmandu, 1988.

APROSC. Evaluation of Training Programmes under Small Farmer Development Programme. 1988.

APROSC. Ongoing Evaluation Study: Dhading District Development Project. 1988.

APROSC. Socio-economic Study on the Integrated Rural Development Project in Lumbini Zone. Kathmandu, 1988.

APROSC. Generation of Employment by the Small Farmer Development Project. Final Report. February 1989.

APROSC. Impact Evaluation - Intensive Banking Programme (IBP), Draft Report, July 1989.

APROSC. Mid-term Evaluation of Small Farmers Development Program - II, 1989.

Asian Development Bank. Appraisal of the Livestock Development Project in Nepal. November 1979.

Asian Development Bank. Project Completion Report of the Second Agriculture Credit Project. October 1980.

Asian Development Bank. Appraisal of the Fifth Agriculture Credit Project in Nepal. March 1987.

Asian Development Bank. Project Completion Report of the Livestock Development Project (Loan No. 445-NEP (SE0)). August 1988.

Bhattarai, Udhav Raj. A Socio-economic Baseline Survey Report on Khairmara Khola Irrigation Project Command Area. Kathmandu, 1986.

Bhattarai, Udhav Raj. A Socio-economic Baseline Survey Report on Mechi Khola Irrigation Project. Kathmandu, 1986.

Bhattarai, Udhav Raj. A Socio-economic Baseline Survey Report on Paini Irrigation Project. SFDP-ADB/Nepal-CARE. Kathmandu, 1986.

Bhattarai, Meera. "Progress Report of Women's Skill Development Project: Nepal Women's Organization," Mahila Bolchhin (Women Speak), Vol XXI, No. 2. 1983.

Central Panchayat Training Institute (CPTI). Mahakali Hills Integrated Rural Development Project, Baseline Survey, Kathmandu, 1984.

Centre for Women and Development. An Evaluation of the Women's Development Project under Small Farmers Development Project. October 1986.

Church, Mary and Singh, S.L. A Mid-Term Evaluation of the Production Credit for Rural Women Project. November 1985.

Cottage and Small Industries Project (CSI) NRB. A Study on the Defaults of Subloan Under CSI Project. 1985.

Cottage and Small Industries Project (CSI) NRB. Credit Position of CSI Project for the Period Ending Mid-July 1988.

HMG/UNICEF/WHO. Joint Nutrition Support Programme in Nepal: Plan of Operations and Recommended Plans of Action. Kathmandu, 1984.

IFAD. Production Credit for Rural Women - Appraisal Report. October 1987.

IFAD. Completion Evaluation Report of the Small Farmers Development Project. May 1989.

IFAD. Production Credit for Rural Women - Appraisal Report. October 1987.

Integrated Development Systems. <u>Integrated Hill Development Project: An Evaluation</u>. Kathmandu, 1982.

Integrated Development Systems. <u>Mid-term Review in the Areas of Women's Income Generation and NGO Programmes</u>, Report submitted to UNFPA, Kathmandu, 1982.

Integrated Development Systems. <u>1982 Semi-Annual Progress Report No. 3 of the Income Generation Promotion Programme (IFFP), Micro-Projects</u> (July 1985 - January 1986), Tansen, Nepal.

Integrated Development Systems. <u>Third Small Farmers Development Project, Nepal, Final Report</u>, submitted to Asian Development Bank and Agricultural Development Bank of Nepal, 1989.

Mitchnik, David. <u>Nepal Rural Development I - Evaluation of the Effects of the Project Credit Component on Beneficiaries</u>. World Bank. Washington, May 1980.

Nepal Food Corporation. <u>NFC: An Overview</u>. Magh, 2045. 1988.

New ERA. <u>A Baseline Survey for the Joint Nutrition Support Programme</u>. Kathmandu, 1986.

New ERA. <u>An Evaluation of the Information, Education and Communication Programme in Nepal</u>. May 1986.

Nippon Koei Co. Ltd. <u>Feasibility Study of Irrigation Development in the Terai Plain (Phase II) - Nepal</u>. Final Report, prepared by FAO acting as an executing agency for UNDP. Tokyo, 1972.

No-Frills Consultants and WPI Inc. <u>Mid-term Evaluation Report, Agricultural Research and Production Project</u> (Project No. 367-0149), January 8, 1988.

Pradhan, B.B. <u>Integrated Rural Development Projects in Nepal: A Review</u>, ICIMOD Occasional Paper No. 2. Kathmandu, 1985.

Puri, Singharaj, Bishnu, and Uprety. <u>Marketing Needs of Women Farmers Experience with PCRW</u>. Kathmandu, 1987.

Saha, Jaya Singh. <u>Impact of the Small Farmers Development Programme on Small Farmers in Nawalparasi</u>. HMG-USAID-GTZ-Winrock Project. 1985.

Shrestha, Behari K. <u>Integration and Coordination of Integrated Rural Development Projects - An Experience from Nepal</u>. Kathmandu: APROSC, CIRDAP, March 1982.

Sir M. Macdonald and Partners Ltd. <u>Medium Irrigation Project, Final Report/Main Report</u>. Hunting Technological Services Ltd., England. 1984.

Stiller, Ludwig. F. <u>Integrated Rural Development Projects in Nepal - An Analysis Based on Project Evaluation Reports</u>. Canadian Co-operation Office. February 1989.

USAID. <u>Rapti Development Project, Economic Analysis</u>. 1986.

World Bank. <u>Project Completion Report: Nepal - Cottage and Small Industries Project</u> (Credit No. 1191-NEP). April 14, 1989.

World Bank. <u>Nepal: Second Cottage and Small Industries Project</u>, Staff Appraisal Report, 1986.

Young, Beverly. "The Primary Education Project," <u>Education and Development</u>, Kathmandu: CERID, Tribhuvan University, 1986.

Distributors of World Bank Publications

ARGENTINA
Carlos Hirsch, SRL
Galería Guemes
Florida 165, 4th Floor-Ofc. 453/465
1333 Buenos Aires

AUSTRALIA, PAPUA NEW GUINEA, FIJI, SOLOMON ISLANDS, VANUATU, AND WESTERN SAMOA
D.A. Books & Journals
648 Whitehorse Road
Mitcham 3132
Victoria

AUSTRIA
Gerold and Co.
Graben 31
A-1011 Wien

BAHRAIN
Bahrain Research and Consultancy
   Associates Ltd.
P.O. Box 22103
Manama Town 317

BANGLADESH
Micro Industries Development
   Assistance Society (MIDAS)
House 5, Road 16
Dhanmondi R/Area
Dhaka 1209

   Branch offices:
   156, Nur Ahmed Sarak
   Chittagong 4000

   76, K.D.A. Avenue
   Kulna

BELGIUM
Jean De Lannoy
Av. du Roi 202
1060 Brussels

CANADA
Le Diffuseur
C.P. 85, 1501B rue Ampère
Boucherville, Québec
J4B 5E6

CHINA
China Financial & Economic Publishing
   House
8, Da Fo Si Dong Jie
Beijing

COLOMBIA
Infoenlace Ltda.
Apartado Aereo 34270
Bogota D.E.

COTE D'IVOIRE
Centre d'Edition et de Diffusion
   Africaines (CEDA)
04 B.P. 541
Abidjan 04 Plateau

CYPRUS
MEMRB Information Services
P.O. Box 2098
Nicosia

DENMARK
SamfundsLitteratur
Rosenoerns Allé 11
DK-1970 Frederiksberg C

DOMINICAN REPUBLIC
Editora Taller, C. por A.
Restauración e Isabel la Católica 309
Apartado Postal 2190
Santo Domingo

EL SALVADOR
Fusades
Avenida Manuel Enrique Araujo #3530
Edificio SISA, 1er. Piso
San Salvador

EGYPT, ARAB REPUBLIC OF
Al Ahram
Al Galaa Street
Cairo

The Middle East Observer
8 Chawarbi Street
Cairo

FINLAND
Akateeminen Kirjakauppa
P.O. Box 128
SF-00101
Helsinki 10

FRANCE
World Bank Publications
66, avenue d'Iéna
75116 Paris

GERMANY, FEDERAL REPUBLIC OF
UNO-Verlag
Poppelsdorfer Allee 55
D-5300 Bonn 1

GREECE
KEME
24, Ippodamou Street Platia Plastiras
Athens-11635

GUATEMALA
Librerias Piedra Santa
5a. Calle 7-55
Zona 1
Guatemala City

HONG KONG, MACAO
Asia 2000 Ltd.
6 Fl., 146 Prince Edward
   Road, W.
Kowloon
Hong Kong

INDIA
Allied Publishers Private Ltd.
751 Mount Road
Madras - 600 002

   Branch offices:
   15 J.N. Heredia Marg
   Ballard Estate
   Bombay - 400 038

   13/14 Asaf Ali Road
   New Delhi - 110 002

   17 Chittaranjan Avenue
   Calcutta - 700 072

   Jayadeva Hostel Building
   5th Main Road Gandhinagar
   Bangalore - 560 009

   3-5-1129 Kachiguda Cross Road
   Hyderabad - 500 027

   Prarthana Flats, 2nd Floor
   Near Thakore Baug, Navrangpura
   Ahmedabad - 380 009

   Patiala House
   16-A Ashok Marg
   Lucknow - 226 001

INDONESIA
Pt. Indira Limited
Jl. Sam Ratulangi 37
P.O. Box 181
Jakarta Pusat

ITALY
Licosa Commissionaria Sansoni SPA
Via Benedetto Fortini, 120/10
Casella Postale 552
50125 Florence

JAPAN
Eastern Book Service
37-3, Hongo 3-Chome, Bunkyo-ku 113
Tokyo

KENYA
Africa Book Service (E.A.) Ltd.
P.O. Box 45245
Nairobi

KOREA, REPUBLIC OF
Pan Korea Book Corporation
P.O. Box 101, Kwangwhamun
Seoul

KUWAIT
MEMRB Information Services
P.O. Box 5465

MALAYSIA
University of Malaya Cooperative
   Bookshop, Limited
P.O. Box 1127, Jalan Pantai Baru
Kuala Lumpur

MEXICO
INFOTEC
Apartado Postal 22-860
14060 Tlalpan, Mexico D.F.

MOROCCO
Société d'Etudes Marketing Marocaine
12 rue Mozart, Bd. d'Anfa
Casablanca

NETHERLANDS
InOr-Publikaties b.v.
P.O. Box 14
7240 BA Lochem

NEW ZEALAND
Hills Library and Information Service
Private Bag
New Market
Auckland

NIGERIA
University Press Limited
Three Crowns Building Jericho
Private Mail Bag 5095
Ibadan

NORWAY
Narvesen Information Center
Book Department
P.O. Box 6125 Etterstad
N-0602 Oslo 6

OMAN
MEMRB Information Services
P.O. Box 1613, Seeb Airport
Muscat

PAKISTAN
Mirza Book Agency
65, Shahrah-e-Quaid-e-Azam
P.O. Box No. 729
Lahore 3

PERU
Editorial Desarrollo SA
Apartado 3824
Lima

PHILIPPINES
International Book Center
Fifth Floor, Filipinas Life Building
Ayala Avenue, Makati
Metro Manila

POLAND
ORPAN
Palac Kultury i Nauki
00-901 Warszawa

PORTUGAL
Livraria Portugal
Rua Do Carmo 70-74
1200 Lisbon

SAUDI ARABIA, QATAR
Jarir Book Store
P.O. Box 3196
Riyadh 11471

MEMRB Information Services
   Branch offices:
   Al Alsa Street
   Al Dahna Center
   First Floor
   P.O. Box 7188
   Riyadh

   Haji Abdullah Alireza Building
   King Khaled Street
   P.O. Box 3969
   Dammam

   33, Mohammed Hassan Awad Street
   P.O. Box 5978
   Jeddah

SINGAPORE, TAIWAN, MYANMAR, BRUNEI
Information Publications
   Private, Ltd.
02-06 1st Fl., Pei-Fu Industrial
   Bldg.
24 New Industrial Road
Singapore 1953

SOUTH AFRICA, BOTSWANA
For single titles:
Oxford University Press Southern
   Africa
P.O. Box 1141
Cape Town 8000

For subscription orders:
International Subscription Service
P.O. Box 41095
Craighall
Johannesburg 2024

SPAIN
Mundi-Prensa Libros, S.A.
Castello 37
28001 Madrid

Libreria Internacional AEDOS
Consell de Cent, 391
08009 Barcelona

SRI LANKA AND THE MALDIVES
Lake House Bookshop
P.O. Box 244
100, Sir Chittampalam A. Gardiner
   Mawatha
Colombo 2

SWEDEN
For single titles:
Fritzes Fackboksforetaget
Regeringsgatan 12, Box 16356
S-103 27 Stockholm

For subscription orders:
Wennergren-Williams AB
Box 30004
S-104 25 Stockholm

SWITZERLAND
For single titles:
Librairie Payot
6, rue Grenus
Case postale 381
CH 1211 Geneva 11

For subscription orders:
Librairie Payot
Service des Abonnements
Case postale 3312
CH 1002 Lausanne

TANZANIA
Oxford University Press
P.O. Box 5299
Dar es Salaam

THAILAND
Central Department Store
306 Silom Road
Bangkok

TRINIDAD & TOBAGO, ANTIGUA BARBUDA, BARBADOS, DOMINICA, GRENADA, GUYANA, JAMAICA, MONTSERRAT, ST. KITTS & NEVIS, ST. LUCIA, ST. VINCENT & GRENADINES
Systematics Studies Unit
#9 Watts Street
Curepe
Trinidad, West Indies

UNITED ARAB EMIRATES
MEMRB Gulf Co.
P.O. Box 6097
Sharjah

UNITED KINGDOM
Microinfo Ltd.
P.O. Box 3
Alton, Hampshire GU34 2PG
England

VENEZUELA
Libreria del Este
Aptdo. 60.337
Caracas 1060-A

YUGOSLAVIA
Jugoslovenska Knjiga
P.O. Box 36
Trg Republike
YU-11000 Belgrade